FINANCIAL ASTROLOGY

Almanac 2023

Trading & Investing Using the Planets

M.G. Bucholtz, B.Sc, MBA, M.Sc.

A WOOD DRAGON BOOK

Financial Astrology Almanac 2023
Trading & Investing Using the Planets

Copyright © 2022 by M.G. Bucholtz

Published by :
Wood Dragon Books,
Box 429, Mossbank, Saskatchewan, Canada, S0H 3G0
http://www.wooddragonbooks.com

ISBN: 978-1-990863-08-0 (Paperback)
ISBN : 978-1-990863-09-7 (eBook)

Contact the author at:
supercyclereport@gmail.com

DEDICATION

To the many traders and investors who, at some visceral level, suspect there is more to the financial market system than P/E ratios and analyst recommendations.

You are correct. There is more. Much more. Rooted in astronomical and astrological timing, the markets are a rich tapestry of interwoven cycles. This book will add a whole new dimension to your trading and investing activities.

DISCLAIMER

All material provided herein is based on material gleaned from mathematical and astrological publications researched by the author to supplement his own trading. This publication is written for those who actively trade and invest in the financial markets and who are looking to incorporate astrological phenomena and esoteric math into their market activity. While the material presented herein has proven reliable to the author in his personal trading and investing activity, there is no guarantee this material will continue to be reliable into the future.

The author and publisher assume no liability whatsoever for any investment or trading decisions made by readers of this book. The reader alone is responsible for all trading and investment outcomes and is further advised not to exceed his or her risk tolerances when trading or investing in the financial markets.

TABLE OF CONTENTS

INTRODUCTION

Many traders and investors think company media releases, media news opinions, quarterly earnings reports, and analyst targets drive stock prices and major index movements.

I disagree. I believe price action on the financial markets is driven by heliocentric and geocentric cycles of planetary movement, by planetary declination, by lunar declination, by angular aspects between planets, and by New Moon planetary aspects. These planetary phenomena affect the emotions of people. Positive emotions create the desire to buy financial instruments. Negative emotions lead to fear and withdrawal. Overlap and interweave together these phenomena and the result will be the ups and downs of price that characterize a stock chart, a commodity price chart, or the chart of a major index. The average trader or investor who remains fixated on media releases and analyst opinions will be unable to discern this rich tapestry of planetary influence.

A question that must be dispensed with is the title of this publication. In particular: where does the term 'astrology' fit into the mix? Webster's dictionary states that: *astronomy is concerned with the study of objects outside the earth's atmosphere.* Webster's further says that: a*strology is the divination of how planets influence our lives.* What this Almanac entails is a blend of astronomy and astrology. This Almanac focuses on planets, planetary aspects, and planetary cycles. These planetary events influence human emotion. Human emotion, in turn, influences price action on financial markets.

A long cycle of planetary activity that overlaps and interweaves with time is the Jupiter/Saturn Gann Master Cycle which unfolds over two decades of heliocentric planetary movement. Threaded through the fabric of this Master Cycle is another long cycle, the McWhirter 18.6-year cycle which aligns to the movement of the North Node through the signs of the zodiac. This nodal cycle broadly defines overall economic activity. Along the way, nodal aspects to major outer planets can exacerbate this economic activity.

The repeated cyclical movement of planets above and below the ecliptic plane aligns to swing highs and swing lows on commodity futures and equity indices.

Irregular occurring events of Mercury or Venus being retrograde often align to trend reversals on stocks and equity indices. A similar observation holds for both Mercury and Venus being at elongation extremes, Superior conjunctions, and Inferior conjunctions.

Cycles also arise from annual celebratory events delineated in the Hebrew lunar-based calendar, notably the seven-year Shemitah cycles. Cycles arise from the mathematics of the Hebrew Alef Bet, and when expressed through the lens of heliocentric movements of Venus or Mars, can be seen aligning to market turning points.

The cycle from one New Moon to the next can also be seen to have a bearing on the market, especially when the Moon transits past 14 degrees of Cancer and 24 degrees of Pisces. Times when the Moon is Void of Course also align to expressions of notable volatility on equity indices.

Even though the mainstream media refuses to embrace astrology or planetary cycles as valid tools for timing the markets, it is my opinion that there are powerful players in the major financial centres of the globe who *do* embrace planetary cycles. Knowing that planetary aspects and cycles influence human emotion, these power players use these occurrences to their advantage to make money in up-trending markets. They also use these cyclic events to induce trend changes on markets. As the price trend turns and markets start to fall, these players profit from their short positions while the average investor on the street experiences emotional angst and sleepless nights knowing the markets are trending down and working against them. "Who are the powerful players that use astrology to move markets?" is a question that burns in my mind. Is it a select group at J.P. Morgan? Is it a group in a dark-panelled office in London? I will likely never know.

"How and why these various cycles have come to be?" is another burning question that remains unanswered. As I have studied these cycles over the past dozen years and applied them to the financial markets, I have developed a new sense of awe for what I deem to be a higher power that guides the Universe. In preparing this manuscript, I had the good fortune to meet a university professor whose research focuses on the Cree First Nations people in Alberta, Canada. He related to me a Cree mythological story involving the Pleiades star cluster which is visible in the constellation Taurus. One day, Sky Woman spotted a far-away planet and expressed a desire to visit it. Spider Woman, who lived amongst the stars of the Pleiades, spun a web so that Sky Woman could reach the far-off planet. The far-off planet was Earth. Mankind originates with Sky Woman and her visit to Earth. This story added to my belief that mankind is hard-wired to events in the cosmos. I was humbled when I subsequently studied stock market reactions during past events of Sun, Venus, and Mars transiting conjunct to the Pleiades star cluster in Taurus. After reading this Almanac, you too may have reason to pause and ponder the power of the cosmos. You may well find yourself feeling humbled.

Paying attention to planetary cycles is not a new concept. Ancient civilizations as far back as the Babylonians recognized planetary cyclical activity, but in a more rudimentary form. Their high priests tracked and recorded changes in the emotions of the people. These diviners and seers also tracked events, both fortuitous and disastrous. Although they lacked the ability to fully comprehend the celestial mechanics of the planetary system, they were able to visually spot the planets Mercury, Venus, Mars, Jupiter, and Saturn in the heavens. They correlated changes in human emotion and societal events to these planets. They assigned to these planets the names of the various deities revered by the people. They identified and named various star constellations in the heavens and divided the heavens into twelve signs. This was the birth of *astrology* as we know it today.

Stories of traders benefiting from planetary activity are also not new. In the early 1900s, esoteric thinkers such as the famous Wall Street trader W.D. Gann reportedly made massive gains when he realized that cycles of astrology bore a striking correlation to financial market action. Gann is most famous for identifying the Saturn/Jupiter cycle which he labelled the Gann Master Cycle. He followed the cyclical activity of Jupiter and Neptune when he traded wheat and corn futures. He also delved deep into esoteric math, notably square root math which led him to develop his Square of Nine. The concept of price squaring with time is also a Gann construct. Today many traders and investors attempt to emulate Gann but they do so in a linear fashion, looking for repetitive cycles on the calendar. What they are missing is the astrology component, which is anything but linear.

In the 1930s, Louise McWhirter greatly illuminated the connection between the stock market and planetary cycles. She identified an 18.6-year cyclical correlation between the general state of the American economy and the position of the North Node of the Moon in the zodiac. Her methodology extended to include the transiting Moon passing by key points of the 1792 natal birth horoscope of the New York Stock Exchange. She also identified a correlation between price movement of a stock and those times when transiting Sun, Mars, Jupiter and Saturn made hard aspects to the natal Sun position in the stock's natal birth (first trade) horoscope. [1]

The late 1940s saw planetary mathematical modelling applied to the stock market when astrologer Garth Allen (a.k.a. Donald Bradley) created his *Siderograph Model* based on aspects between the various transiting planets. Each aspect as it occurs is given a sinusoidal weighting as the orb (separation) between the planets varies. Bradley's model was obscured in the aftermath of the 2008 financial crisis when the Federal Reserve was injecting massive amounts of liquidity into the financial system. Now that the Fed has embarked on a monetary tightening policy, Bradley's model is again proving itself a powerful indicator of trend changes on the S&P 500. (2)

As the 1950s dawned, academics at institutions like Yale and Harvard came to dominate discussions of the financial markets. Talk of planetary cycles influencing financial markets was soon swept aside out of public view. Cyclical analysis was replaced by academic constructs like *Modern Portfolio Theory* and the *Efficient Market Hypothesis*. These persisted for several decades until coming under severe scrutiny with the 2000 tech bubble meltdown and again with the 2008 sub-prime mortgage crisis which nearly derailed the global economy.

In the past decade, the application of planetary science to the stock market has been elevated and made more user friendly. The software designers at *Market Analyst/Optuma* now have an impressive financial astrology platform built into their charting program. More recently, author and trader Fabio Oreste published a book entitled *Quantum Trading* in which he describes *quantum lines*, a mathematical construct based on the work of Einstein, Niels Bohr, and Bernhard Reimann. Oreste shows how quantum lines can be used to delineate areas of price support and resistance.

You have probably experienced the effects of the planets on the financial market without even realizing it. Think back to the dark days of late 2008 when there was genuine concern over the very survival of the financial market system. This timeframe was the end of an 18.6-year cycle of the North Node traveling around the zodiac. To high-level, power players in the financial system who understood astrology, this period was a prime opportunity to feast off the fear of the investing public and the anxiety of government officials who were standing at the ready with lucrative bailout packages. The market low in March 2009 came at a confluence of a Mars and a Neptune quantum point. Curiously enough, Mars and Neptune are deemed to be the planetary rulers of the New York Stock Exchange. The March 2009 low also aligned perfectly to the start of Venus being retrograde.

Think back to August 2015 and the market selloff that the financial media did not see coming. This selloff started at a confluence of three events: Venus being retrograde, the appearance of Venus as a Morning Star after having been only visible as an Evening Star for the previous 263 days, and the close conjunction of Venus and Jupiter (a 24-year cyclical event).

Remember the early days of 2016 when Mercury was retrograde and the markets hit a rough patch? Remember the weakness of June 2016 when Venus emerged from conjunction to become visible as an Evening Star?

Do you recall the dire predictions for financial market calamity following the 2016 election of Donald Trump to the White House? When the markets instead powered higher, analysts were flummoxed. Venus was making its declination minima right at the time of the American election. Venus declination minima events bear a striking correlation to changes of trend on US equity markets.

What about the early days of 2018 when fear once again gripped the system? Venus was at its declination minimum. Markets reached another turning point in the first week of October, 2018 when Venus was again at a declination low. Add the fact that Venus turned retrograde at the same time and the fear starts to make sense. Markets sold off sharply into mid-December before starting to recover. Sun was conjunct Saturn at this time which correlates strongly to trend changes on equity markets. The North Node had also just changed zodiac signs, an event which also aligns with trend changes.

Markets hit a sudden rough patch in early August 2019. Mercury had just finished retrograde; Moon transited a key point on the NYSE

1792 natal horoscope; and the Federal Reserve cut interest rates due to overnight repurchase agreement (repo) market liquidity concerns.

US equity markets peaked in late February 2020 and went into total spasm in March 2020. Mercury was retrograde and Mars had just made its declination minimum. Venus was at the same degree of declination it had been in 1792 when the NYSE was founded. In addition, a powerful fractal cycle from a crisis event in January 1920 suggested another crisis event in March 2020. COVID would soon become a new word in our vocabulary.

Do you recall the confusion surrounding the 2020 election of Joe Biden to the White House? The day of the election, Mercury finished retrograde and heliocentric Jupiter and Saturn were exactly at 0-degrees of separation. To have these two events occur right on the election date is a rarity. The events concerning the validity of this election will be hotly debated for years to come.

The events of late 2021 reminded us all again of the power of planetary events. As 2021 was ending, the equity markets were peaking. Mercury was approaching its greatest easterly elongation. Venus was retrograde and also was approaching its Inferior Conjunction. Mercury would soon turn retrograde. Mars was approaching its declination minimum. This was a concentration of cosmic energy at work. The powerful players, whoever they were, took full advantage. They pushed the markets into a downtrend through aggressive selling. Individual investors whose emotions were rattled by the cosmic events panicked and started to feed into the downtrend. By June 2022, the selling had caused a full 20% decline across equity markets.

The period September 2021 to September 2022 was a Shemitah Year in the Hebrew faith (a one-in-seven-year occurrence). I predicted at the outset that this Shemitah Year would deliver some headline events. And sure enough, Russia invaded Ukraine, inflation surged, gasoline prices at the pump jumped higher, and Europe stumbled into an energy crisis.

And, so it goes. Cycles continue to unfold as time marches on. People who view the markets through the lens of analyst opinions and media blather will be unable to see these cycles, which are hidden in plain view. They will ride an emotional roller coaster as their financial planners tell them that investing is for the long term and not to worry. On the other hand, investors who are able to identify these cycles will be able to take steps to protect themselves and profit accordingly.

When I began to embrace financial astrology in 2012, it was a monumental shift for me. My educational background comprises an Engineering degree, an MBA degree and a M.Sc. degree. My approach to the financial market and to the planets is thus heavily slanted towards mathematics and analytical science.

This Almanac begins by offering the reader a look at the basic science of astrology. What then follows is an examination of the various planetary cycles that I believe drive the performance of the financial markets. For each type of cycle, I delineate times in the calendar months of 2023 when investors ought to be alert to possible trend changes on the US equity markets. My focus is on the US equity markets because in this age of global connectedness, moves on the S&P 500 are often quickly reflected in other global indices. I then provide a look at various commodity futures and the astro phenomena that influence price action. I delineate times when traders ought to be alert to possible trend changes. I conclude with offering a calculated list of quantum price lines for major indices and commodity futures. Along the way, I offer insight into fractal patterns, Kabbalah mathematical patterns, and planetary declination.

When applying planetary cycles and astrology to trading and investing, it is vital at all times to be aware of the price trend. The Parabolic Stop & Reverse and Volatility Stop chart indicators developed by J. Welles Wilder are very effective at identifying trend changes. The Ergodic Oscillator developed by William Blau in the 1980s is a powerful trend indicator. In addition, the chart indicators created by Martin Pring and George Lane are also potent. In this edition of the Almanac, I have added a chapter to further explain how to use these various chart indicators.

Using the planets for financial investing is not about taking action at each and every astro event that comes along, because not all astro events are powerful enough to induce a change of trend. Instead, look for a change of trend that aligns to a planetary event. When you see the trend change, you should take action. Whether that action is implementing a long position, a short position, an options strategy, or just tightening up on a stop loss will depend on your personal appetite for risk, and your investment and trading objectives.

Author Note: This Almanac, which is my tenth such annual publication, is designed to be a resource to help you stay abreast of the various astro events that 2023 holds in store. I am also the author of several other astrology books and publish a bi-weekly subscription-based newsletter called *The Astrology Letter* and a monthly subscription-based newsletter called *The Cycle Report*. Through all of my written efforts, I hope to encourage people to embrace financial astrology as a valuable tool to aid in trading and investing decision making.

CHAPTER ONE
Fundamentals

The Sun is at the center of our solar system. The Earth, Moon, planets and other asteroid bodies complete the planetary system. In addition to the Sun and Moon, there are eight celestial bodies important to the application of planetary cycles to the financial markets. These planets are Mercury, Venus, Mars, Jupiter, Saturn, Uranus, Neptune, and Pluto. Figure 1-1 illustrates these various bodies and their spatial relation to the Sun. Mercury is the closest to the Sun while Pluto is the farthest away.

The Ecliptic and the Zodiac

The various planets and other asteroid bodies rotate 360-degrees around the Sun following a path called the *ecliptic plane*. As shown in Figure 1-2, Earth and its Equator are slightly tilted (approximately 23.45-degrees) relative to the ecliptic plane.

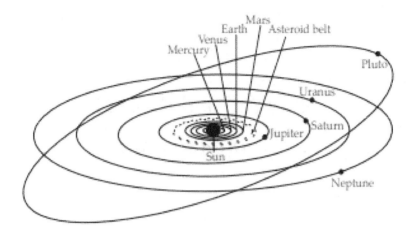

Figure 1-1
The Planets

Projecting the Earth's equator into space produces the *celestial equator plane*. There are two points of intersection between the ecliptic plane and celestial equator plane. Mathematically, this makes sense as two non-parallel planes must intersect at two points. These points are commonly called the *vernal equinox* (occurring at March 20th) and the *autumnal equinox* (occurring at September 20th). You will recognize these dates as the first day of Spring and the first day of Fall, respectively. Dividing the ecliptic plane into twelve equal sections of 30-degrees results in what astrologers call the *zodiac*. The twelve portions of the zodiac have names such as Aries, Cancer, and Leo. Ancient civilizations looking skyward identified patterns of stars called constellations that aligned to these twelve zodiac divisions. (If these names sound familiar, they should. You routinely see all twelve names in the daily horoscope section of your morning newspaper).

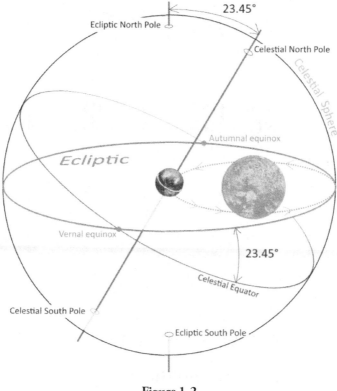

Figure 1-2
The Ecliptic

The Glyphs

Figure 1-3 illustrates the symbols that appear in the twelve segments of a zodiac wheel. The segments are more properly called *signs*; the symbols are called *glyphs*.

The starting point or zero degree point of the zodiac wheel occurs in the sign Aries at the vernal equinox. The vernal equinox is when, from our vantage point on Earth, the Sun appears at 0-degrees Aries. The autumnal equinox is when, from our vantage point on Earth, the Sun appears at 180-degrees from 0-degrees Aries (0-degrees of Libra).

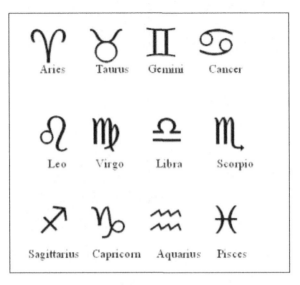

Figure 1-3
The Zodiac Wheel

The various planets are also denoted by glyphs, as shown in Figure 1-4.

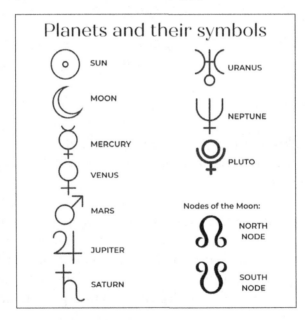

Figure 1-4
The Glyphs

Declination

As the various celestial bodies make their respective journeys around the Sun, they can be seen to move above and below the ecliptic plane. This movement is termed *declination*. Celestial bodies experience declinations of up to about 25-degrees above and below the ecliptic plane. Declination occurs as a result of the force of gravitational pull on a planet. The larger outer planets (Jupiter, Saturn, Neptune, Uranus and Pluto), owing to their size and distance from the Sun experience declination changes that are slower to evolve. The smaller inner planets (Mercury, Venus, Mars) exhibit more amplitude in their declination patterns.

As this Almanac will illustrate, changes in the declination of a celestial body (most notably Mars and Venus) can affect the financial markets. W.D. Gann believed that the dates Venus and Mars return to the same declination level they were at when a stock or a commodity future first started trading (first trade date/natal declination level) can align to price trend changes.

In addition to planetary declination, this Almanac will also reference the declination of Sun. Of course, the Sun is not moving up and down. Rather, as planet Earth orbits the Sun, Earth exhibits a change in the tilt relative to its axis. This affects the angle of the Sun's rays striking the Earth. Astronomers have thus adopted the standard approach of referring to the declination of the Sun, not to the declination of the Earth.

The Moon

Just as the planets orbit 360-degrees around the Sun, the Moon orbits 360-degrees around the Earth. The Moon orbits the Earth in a plane of motion called the *lunar ecliptic plane*. This plane is inclined at about 5-degrees to the ecliptic plane as Figure 1-5 shows. The Moon orbits Earth with a slight elliptical pattern in approximately 27.3 days, relative to an observer located on a fixed frame of reference such as the Sun. This time period is known as a *sidereal month*. However, during one sidereal month, an observer located on Earth (a moving frame of reference) will revolve part way around the Sun. Because of this added movement, the Earth-bound observer will see a complete orbit of the Moon around the Earth in approximately 29.5 days. This 29.5-day period of time is known as a *synodic month* or more commonly a *lunar* month. The lunar month plays a key role in discerning the volatility of the financial markets.

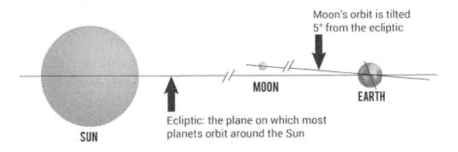

Figure 1-5
Lunar Orbit

The Nodes

A mathematical construct related to the Moon, and central to financial markets, is the *Nodes*. The Nodes are the points of intersection between the Earth's ecliptic plane and the Moon's ecliptic plane. In astrology, typically only the North Node is referred to. The North Node forms the basis for the McWhirter Method which will be discussed in Chapter 3.

Synodic and Sidereal Cycles

The concepts of synodic and sidereal also extend beyond the Moon to include all the planets. To an earth-bound observer, a synodic time period is the time between two successive planetary occurrences. That is, how many days does it take for Sun passing Pluto on the zodiac wheel to Sun again passing Pluto? To a Sun-bound observer (a fixed frame of reference), a sidereal time period is the number of days (or years) it takes for a planet to orbit the Sun. The table in Figure 1-6 presents synodic and sidereal data.

PLANET	SYNODIC PERIOD	SIDEREAL PERIOD
Mercury	116 days	88 days
Venus	584 days	225 days
Mars	780 days	1.9 years
Jupiter	399 days	11.9 years
Saturn	378 days	29.5 years
Uranus	370 days	84 years
Neptune	368 days	164.8 years
Pluto	367 days	248.5 years

Figure 1-6
Synodic and Sidereal Data

Geocentric and Heliocentric Astrology

The terms synodic and sidereal help define the two distinct varieties of astrology – geocentric and heliocentric.

In *geocentric* astrology (synodic), the Earth is the vantage point for observing the planets as they pass through the signs of the zodiac.

Owing to the different times for the planets to each orbit the Sun, an observer situated on Earth will see the planets making distinct angles (called *aspects*) with one another and also with the Sun. The aspects that are commonly used in astrology are 0, 30, 45, 60, 90, 120, 150 and 180-degrees. In financial astrology, it is common to refer to only the 0, 90, 120 and 180-degree aspects.

In *heliocentric* astrology (sidereal), the Sun is the vantage point for observing the planets as they pass through the signs of the zodiac. An observer positioned on the Sun would also see the orbiting planets making aspects with one another.

To identify these aspects, astrologers use Ephemeris tables. For geocentric astrology, the *New American Ephemeris for the 21ˢᵗ Century* is commonly used. For heliocentric astrology, the *American Heliocentric Ephemeris* is a good resource.

Instead of a book of tabular data, quicker aspect determination can be made with software. Two excellent software programs available are *Millenium Trax* produced by AIR Software and *Solar Fire Gold* produced by software company Astrolabe. My preference is the Solar Fire Gold product. I also use a market platform called *Optuma/Market Analyst*. This brilliant piece of software, originally developed in Australia, allows the user to generate end-of-day price charts for equities and commodities from a multitude of exchanges and then overlay various planetary aspects and cycle occurrences onto the chart. As your journey into trading and investing using the planets deepens, you might be tempted to spend the money to acquire a software program.

Ascendant, Descendant, MC and IC

As the Earth rotates on its axis once in every 24 hours, an observer situated on Earth will detect an apparent motion of the constellation stars that define the zodiac. To better define this motion, astrologers apply four cardinal points to the zodiac, almost like the north, south, east and west points on a compass. These cardinal points divide the zodiac into four quadrants. The east point is termed the *Ascendant* and is often abbreviated *Asc*. The west point is termed the *Descendant* and is often abbreviated *Dsc*. The south point is termed the *Medium Coeli* (Latin for Mid Heaven) and is often abbreviated *MC* or *MH*. The north point is termed the *Imum Coeli* (Latin for *bottom of the sky*) and is abbreviated *IC*.

Figure 1-7 illustrates the placement of these cardinal points on a typical zodiac wheel. The importance of the Ascendant and Mid-Heaven will be emphasized in more detail when the McWhirter Method is discussed in Chapter 3.

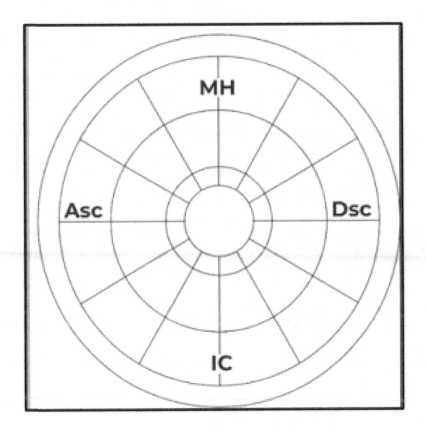

Figure 1-7
Cardinal Points

Retrograde

The term 'retrograde' is taken from the Latin expression *retrogradus* which means 'backward step'.

From our vantage point on Earth, we describe the position of the planets relative to one of the twelve star constellations in the sky. There will be three (occasionally four) times during a year when Earth and Mercury pass by each other (Mercury retrograde). There will be one time (occasionally two times) per year when Earth and Venus pass each other (Venus retrograde). There will be one time every two years when Earth and Mars pass each other (Mars retrograde). As one of these planets starts to lap past Earth, we can see the planet's position relative to one of the star constellations in the sky. Owing to the different orbital speeds of Earth and the lapping planet, there will be a period of time when we see the lapping planet in what appears to be the previous constellation. For example, we might start off seeing Mercury against the star constellation of Gemini. As Mercury begins to lap past Earth, we will see Mercury against the star constellation of Taurus. As Mercury passes by Earth, we will see Mercury again in Taurus. Of course, Mercury has not physically reversed course and moved backwards. This is an optical illusion created by the different orbital speeds of Mercury and Earth.

These brief illusory periods are what astrologers call *retrograde* events. To ancient societies, retrograde events were of great significance as human emotion was often seen to be changeable at these events.

Retrograde events involving these inner planets very often lead to short term price trend changes developing. Is it possible that our DNA is hard-wired such that we feel uncomfortable at retrograde events? Does this emotional discomfort compel us to buy or perhaps sell on the financial markets?

Elongation and Conjunction

From an observer's vantage point on Earth, there will also be times when planets are at maximum angles of separation from the Sun. These events are what astronomers refer to as *maximum easterly* and *maximum westerly* elongations. These events definitely have a correlation to trend changes on the markets.

Mercury and Venus are closer to the Sun than is the Earth. From our vantage point on Earth, there will be times when Mercury and Venus are situated between the Earth and the Sun. There will also be times when the Sun is between the Earth and Mercury or Venus. On the zodiac wheel, the times when Mercury or Venus are in the same zodiac sign and degree as the Earth are what astronomers call *conjunctions*.

An *Inferior Conjunction* occurs when Mercury or Venus is between Earth and the Sun.

A *Superior Conjunction* occurs when the Sun is between Earth and Mercury or Venus. Figure 1-8 illustrates the concept of elongation and conjunction.

Inferior Conjunction events occur on either side of retrograde events. For example, Venus was retrograde from December 19, 2021 through January 28, 2022. Its exact Inferior Conjunction was recorded on January 8, 2022. The peak on the S&P 500 in early January 2022 was directly connected to these Venus phenomena which disturbed human emotion. As unsettled investors began selling, the trend on the S&P 500 changed to negative. The negative sentiment remained intact until mid-June, 2022 when the Sun recorded its declination maximum. Following this brief reprieve, the trend soon shifted back to negative. Such are the intricacies of the cosmos and their effects on human emotion and the markets.

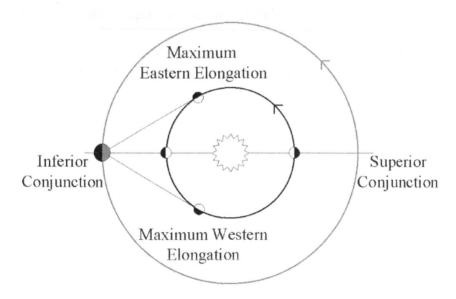

Figure 1-8
Superior and Inferior Conjunction

After Venus has been at Inferior Conjunction, it will be visible as a Morning Star.

After it has been at Superior Conjunction, it will be visible as an Evening Star.

Venus was at Superior Conjunction on March 28, 2013 (8 Aries), October 25, 2014 (1 Scorpio), June 6, 2016 (16 Gemini), January 8, 2018 (18 Capricorn), August 14, 2019 (20 Leo), March 26, 2021 (6 Aries), and October 22, 2022 (28 Libra).

Venus was at Inferior Conjunction on June 6, 2012 (15 Gemini), January 11, 2014 (21 Capricorn), August 15, 2015 (22 Leo), March 25, 2017 (4 Aries), October 26, 2018 (3 Scorpio), June 3, 2020 (14 Gemini), and January 8, 2022 (19 Capricorn).

If one plots consecutive Superior Conjunction events on a zodiac wheel, the plot points can be joined to form a 5-pointed star called a pentagram. Likewise, the Inferior Conjunction events can be plotted and joined to form a pentagram. Such is the elegance and mystique of the cosmos.

Having looked at these basics, let's next engage in a deeper exploration of events of the cosmos and how these events can impact financial markets.

CHAPTER TWO

The Gann Master Cycle

W.D. Gann closely followed the cycles of Jupiter and Saturn. To an observer situated on the fixed (heliocentric) vantage point of the Sun, Jupiter would be seen orbiting the Sun in about 12 years and Saturn in just over 29 years. Gann interpreted these orbital cycles one step further and noted that every 19.86 years, heliocentric Jupiter and Saturn were at conjunction in a particular sign of the zodiac and separated by 0-degrees. This 19.86-year time span is what he called the *Master Cycle*.

The curious feature of the Master Cycle is that the occurrence of the Jupiter/Saturn conjunction does not always align precisely with a change of trend. Consider the following examples:

✿ The market weakness in late 1901 aligned to a conjunction of these two outer planets. But following this conjunction event, the Dow Jones did not reach a definitive turning point low until late 1903.

✪ In 1920, the U.S. economy encountered a recession. In August 1921, Jupiter and Saturn reached conjunction. After conjunction, the Dow Jones started to rally higher, but then severely re-tested the lows in late 1923.

✪ In late 1940, Jupiter and Saturn again were at conjunction. The Dow Jones did not record a definitive low until early 1942.

✪ In April 1961, Jupiter and Saturn again were at conjunction. The Dow Jones reached a turning point low in mid-1962.

✪ In the Spring of 1981, Jupiter and Saturn recorded a conjunction event. It would not be until August 1982 that the Dow Jones recorded a turning point low that evolved into a massive bull market run that endured until the next conjunction event in June 2000.

✪ Following this 2000 conjunction and the start of a new Master Cycle, it would not be until October 2002 that the Dow Jones reached a turning point low.

✪ Figure 2-1 illustrates how the Dow Jones Average was making its peak in mid-2000 as Jupiter and Saturn were making their conjunction.

✪ In November 2020, heliocentric Saturn and Jupiter again made their 0-degree conjunction right at the time of the US Presidential election. In the aftermath of the election, the markets maintained a positive tone. This was due in large measure to the Federal Reserve injecting $120 billion per month of liquidity into the financial system in response to the COVID crisis.

But in a pattern reminiscent of past Jupiter-Saturn conjunctions, the market tipped over into a downtrend many months after the conjunction; this time at the end of 2021. At this time of writing in October 2022, it appears that the market is nearing a turning point low after having declined 25% in 2022 year-to-date.

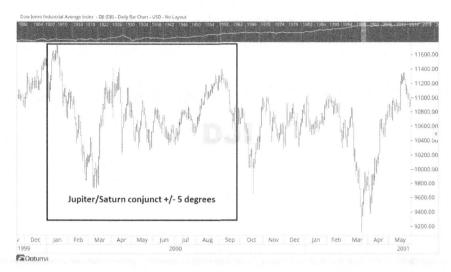

Figure 2-1
Jupiter / Saturn Conjunction in 2000

The time to reach this turning point low has been influenced by a tapering of Federal Reserve liquidity injections, post-COVID global supply chain issues, and inflation running hot. Figure 2-2 illustrates further. Will the equity markets move decidedly higher in the coming years as has often been the case with past Gann Master Cycles? Or, is the post-COVID economy too weak to exhibit higher equity markets?

The recent display of early-stage Master Cycle weakness was also heavily influenced by the Shemitah year. In September 2021, a Shemitah year commenced and it ran until September 2022.

Shemitah years are notorious for unexpected market moves. The events of the 2021-2022 Shemitah were not at all surprising in their scope and magnitude. BitCoin languished. Russia invaded Ukraine. Equity markets swooned. Inflation surged into the high single digits. Oil prices surged which sent gasoline prices at the pumps to uncomfortable levels. Grain prices peaked and then fell. Fertilizer prices surged, placing a new financial burden on farm operators. A new virus called Monkey-Pox made an appearance on the world stage. These events will have lasting effects. Such is the power of a Shemitah year. The concept of a Shemitah year is explained further in Chapter 9.

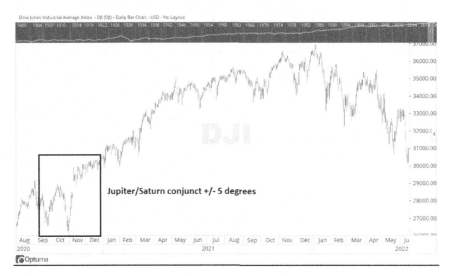

Figure 2-1
Jupiter / Saturn Conjunction in 2020-2022

Triple recurrences of the Gann Master Cycle can have a repetitive impact on geopolitics and on the financial markets. For example, *heliocentric* Jupiter and Saturn were conjunct in April, 1961 in the latter part of Capricorn. The 1981 conjunction occurred in the sign of Libra. The June 2000 conjunction occurred in the sign of Taurus. The November 2020 conjunction occurred at 0-degrees Aquarius, a mere seven degrees from where the 1961 conjunction had occurred. This seems to imply that after a 60 year time span, history could start to rhyme again.

Consider that the US equity market recorded its post-1929 crash low in 1932. Forward 60 years and 1982 marked the start of what would be a major bull market. The year 1937 marked an interim peak on the US stock market. Forward 60 years and the 1997 Asian currency crisis comes into focus. A low was recorded on the market in 1942. Forward 60 years and a similar event occurred. A low was again recorded in 1949. Forward 60 years and the March 2009 lows make an appearance. Late 1961 marked a turning point on the market. Add 60 years and late 2021 likewise presented a turning point. From a low in 1962 the US market rallied briskly into 1965. **This pattern seems to imply that 2023 to 2025 could see a firm market.** This seems hard to fathom at this time of writing, with geopolitical tensions spilling over, inflation running hot, and recessionary fears mounting. However, the 60-year rhyming track record suggests that the unbelievable might occur. To put this all in perspective, one should consider that the US economy, despite its internal political struggles and inflationary issues, is a strong economy. The 60-year rhyming pattern may not give wildly higher markets. Instead, we might see comfortably higher markets, returning perhaps single digits.

CHAPTER THREE
The 18.6 Year Cycle

In addition to the Gann Master Cycle, there exists another long cycle that has a powerful impact on the financial markets (and the global property market as well).

This cycle was first written about in the 1930s by a mysterious figure called Louise McWhirter. (1) I say *mysterious* because in all my research I have neither come across any other writings by her nor have I found reference to her in other manuscripts. I am almost of the opinion that the name was a pseudonym for someone seeking to disseminate astrological ideas while remaining anonymous.

The Moon orbits the Earth in a plane of motion called the lunar ecliptic. Two planes that are not parallel will always intersect at two points. The two points where the lunar ecliptic intersects that plane of motion of planet Earth are termed the *North Node* and *South Node*. The two Nodes are opposite one another in the zodiac wheel. Common practice is to focus only on the North Node.

McWhirter recognized that the transit of the North Node of the Moon around the zodiac wheel takes 18.6 years and that the Node progresses in a backwards (retrograde) motion through the zodiac signs.

Through examination of copious amounts of economic data provided by Leonard P. Ayers of the Cleveland Trust Company, McWhirter was able to conclude that when the North Node moves through certain zodiac signs, the economic business cycle reaches a low point. When the Node is passing certain other signs, the business cycle is at its strongest.

This line of thinking is still with us today. A notable authority embracing this economic cycle is British economist Fred Harrison. In his published works, he discusses this long economic cycle going back to the Industrial Revolution. But, to maintain respect in academia, he stops just shy of stating a connection to the zodiac and the North Node.

McWhirter was able to discern the following from the Cleveland Trust data:

- ☼ As the Node enters Aquarius, the low point of economic activity is reached.

- ☼ As the Node leaves Aquarius and begins to transit through Capricorn and Sagittarius, the economy starts to return to normal.

- ☼ As the Node passes through Scorpio and Libra, the economy is functioning above normal.

- ☼ As the Node transits through Leo, the high point in economic activity is reached.

- ☼ As the Node transits through Cancer and Gemini, the economy is easing back towards normal.

- ☼ As the Node enters the sign of Taurus, the economy begins to slow.

- ☼ As the Node enters Aquarius, the low point of economic activity is reached and a full 18.6 year cycle is completed.

McWhirter further observed some secondary factors that could influence the tenor of economic activity in a good way, regardless of which sign the Node was in at the time:

- ✿ Jupiter being 0-degrees conjunct to the Node

- ✿ Jupiter being in Gemini or Cancer

- ✿ Pluto being at a favorable aspect to the Node

McWhirter also observed some secondary factors that can influence the tenor of economic activity in a bad way, regardless of which sign the Node was in at the time:

- ✿ Saturn being 0, 90, or 180-degrees to the Node

- ✿ Saturn in Gemini or Cancer

- ✿ Uranus in Gemini

- ✿ Uranus being 0, 90 or 180-degrees to the Node

- ✿ Pluto being at an unfavorable aspect to the Node.

In early 2020, the North Node was in the sign of Cancer. The economy was gently easing, in alignment with McWhirter's predictions. None of the above mentioned secondary bad factors were in play.

The Node entered Taurus in early 2022 and the economy started to encounter stiff headwinds. These headwinds were emboldened by strained post-COVID supply chains, rising inflation, a flattening yield curve, and a Russian invasion of Ukraine. Market strategists soon started talking about recession.

Starting in late March 2022, Saturn formed a hard, 90-degree square aspect to the Node. In alignment with McWhirter's findings, this 90-degree square event negatively impacted the already slowing economy. Waves of selling pressure across equity markets lasted into late June. Saturn turned retrograde on June 4, 2022. As it did, the reins of power were handed off to Uranus which then started its march towards a conjunction with Node. The exact conjunction of Uranus to Node occurred on July 30. Once this exact conjunction was complete, Uranus and Node slowly drifted apart, maintaining a 3-4 degree unfavorable separation which still qualified as a conjunction. The negative tenor from this conjunction created another weave of weakness on equity markets that persisted into October, 2022.

At this time of writing in October 2022, the Node is at 13 degrees of Taurus and will enter Aries in mid-2023. Without doubt, the tenor of the economy is slowing. At this time of writing in October 2022, the markets remain weak, but volatile. Talk of recession continues, and inflation remains stubbornly entrenched. Financial analysts continue to forecast a recession, but have mixed opinions on its severity. By late December, Uranus and Node will have pulled farther apart to a 5-degree separation. Talk of recession might abate somewhat as a result.

However, by April 2023, Pluto will be within 3-degrees of a square aspect (unfavorable) to the Node. Economic weakness will likely start to dominate news headlines yet again. The Federal Reserve might be forced to talk about a reversal of course on its monetary policy. Pluto will turn retrograde on May 2, 2023. Whether this weakens or emboldens its impact on the economy remains to be seen. McWhirter did not specifically address retrograde activity in her discussion of secondary factors that influence economic activity. As Pluto turns retrograde, Jupiter will be 7-degrees away from a conjunction (favorable) to the Node. This might start to allay any fears of a recession. On May 25, Jupiter will be at its 0-degree conjunction to the Node. By mid-October, 2023 Pluto will be direct again, but at a 2-degree separation (unfavorable) from the Node. This could temporarily re-kindle talk of economic slowdown. By 2023 year end, Pluto will be 7-degrees apart from the Node and of lesser concern.

By the time the Node enters Pisces in early 2025, the odds of economic trouble will have decidedly increased. By mid-2026, the Node will be in Aquarius to mark the end of the 18.6-year cycle. I will not be surprised to see another financial crisis in late 2026 and into 2027.

There is one more cyclical event that troubles me. I hate to think about it, but it bears discussion. In 2026, Uranus will be in Gemini and also at a hard 90-degrees to the Node. This astro positioning warns of a negative economic time. It also warns of possible war. The 1776 natal horoscope for the USA has Uranus in the sign of Gemini. By mid-2027 Uranus will be exactly conjunct to the 1776 Uranus natal position at 8 degrees of Gemini. Uranus takes 84 years to travel one time through the zodiac. Subtract Uranus cycles from the year 2027 and past dates aligning to World War II, the US Civil War, and the War of Independence all come into focus. One need not look too hard to see how fragile the global geopolitical situation is. The seeds for a 2027 catastrophe may already have been sown.

⚙ **The McWhirter 18.6-year cycle suggests the economy is on a slowing trajectory. Uranus conjunct the North Node in the final months of 2022 will keep recessionary fears stoked.**

⚙ **These fears might diminish in the first part of 2023. As Pluto moves into a 90-degree unfavorable square position in Q2 of 2023, recessionary fears might re-surface.**

⚙ **In late Q2, Pluto will have turned retrograde and Jupiter will have moved into a favorable conjunction to the Node.**

⚙ **In late Q3, Pluto might once again re-kindle recessionary concerns, but by 2023 year end Pluto will no longer be a concern.**

CHAPTER FOUR
Fractal Cycles

An examination of economic data, as well as price movements on stock markets, shows the occurrence of seemingly unexpected crisis events over time. A closer, mathematical study of these events reveals they are not random, nor unexpected. Rather they are often based on fractal patterns and the Fibonacci-related value of 0.618 (phi). *Phi is the inverse of the Golden Ratio of 1.618.* Dividing successive terms of the Fibonacci sequence by their preceding term shows that the results converge to the Golden Ratio of 1.618. The Golden Ratio can be seen in growth patterns and natural phenomena throughout Nature. The Golden Ratio is also often applied to analyzing price movements on the stock market.

A fractal pattern centers around iterations of a basic shape. Consider the shapes shown in Figure 4-1. From the basic triangle shape (called the seed pattern), more complex patterns can be created.

In the financial markets the starting point for a repetitive fractal can be taken from a wide variety of seed events. Late in 2021, while perusing the shelves at a used bookstore I happened upon a book entitled *Fractal Time* written by author Gregg Braden. [1]

35

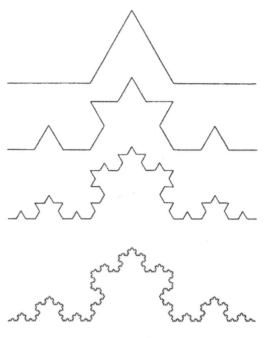

Figure 4-1
Seed Pattern

Braden's starting point for fractal analysis is the *Precession of the Equinox*. In the 3rd week of March each year (around March 20), we celebrate the Spring Equinox. The precise time of the Equinox coincides with the declination of the Sun being at 0-degrees above the ecliptic. Currently, the Spring Equinox occurs with the Sun at 0-degrees in the sign of Aries. Every 71 years, the Equinox will occur 1-degree earlier. After 25,878 years, the Equinox will again be back at 0-degrees of Aries. This backwards movement is the Precession of the Equinox. Braden divides the 25,878-year period into five intervals of 5175 years each. Braden argues that we are currently in the midst of one of these 5175-year periods which began in 3114 BC. The period will end in 2061.

Using Braden's method, take a year that is of curious interest (the seed event) and add to it the value of 3114. Then divide this sum by 5175 and multiply the result by phi (0.618). The result of this multiplication will next be multiplied by the remaining years in the cycle.

For example, in January 1819, there was a financial panic in the US. Author Clyde Haulman, writing in the journal *Financial History*, [2] describes how in the early 1800s many individual states in the western part of America had issued their own bank notes. The eastern part of the US was using bank notes issued by the federal government. The notes from the western states started depreciating in value around 1816. Holders of these western notes attempted to redeem them for cash at eastern banks but were soon stopped by the US Treasury Department. This monetary contraction set the stage for an eventual panic. Haulman writes:

With a monetary contraction underway, the continued retirement of federal debt, much of it to foreigners, and declines in the overseas markets for American staples, the United States economy was headed for disaster. The most dramatic aspect of the disaster was a rapid deflation as prices fell 30.6 percent between 1818 and 1821. Stagnation of real output that for some parts of the country lasted well into the 1820s. Real GNP fell in 1819 and was flat over the period 1818–1821. The young republic's rude introduction to boom-and-bust capitalism reported by these sources was a complex combination of financial market volatility, swings in international market demand, and federal government financial activity.

January 1819 seems to have been the worst of the crisis as bank credit plunged.

Applying Braden's method, January is $1/12^{th}$ of the calendar year. In decimal form, $1/12^{th}$ is 0.083. In each of the following examples, the symbol * denotes the mathematical operation of multiplication.

1819.083 + 3114 = 4933.083
(4933.083/5175) * phi = 0.58911
5175 – 4933.083 = 241.917
241.917 * 0.58911 = 142.515
4933.083 + 142.515 = 5075.59
5075-3114 = 1961.598 = July 1961.

These calculations suggest that a fractal event stemming from the 1819 event would occur in mid-1961.

A look back at history shows that in mid-1961 the stock market was reaching a peak. Stocks had become too expensive and a return to more sensible levels was overdue. A handful of months later, the market starting selling off. The damage ended with a drawdown of 22%. This event has been dubbed the *Kennedy Slide*. When I wrote the manuscript for the 2021 Almanac, I presented this very same example. I further took the 1961.598 figure and used Braden's method to arrive at a fractal value of 2021.848. In round figures this equates to a timeframe in late October 2021. A look back at the late October 2021 period shows that the S&P 500 reached the 4700 level. Following a dip towards 4500 in early December, the S&P 500 vainly tried to rally into year-end but ran out of steam. Once again, fractal math proved accurate in its forecasting ability.

As another example, there was a financial panic event in 1893. Richardson and Sablik [3] of the Federal Reserve Bank of Richmond summarize the panic as follows:

The Panic of 1893 was one of the most severe financial crises in the history of the United States. The crisis started with banks in the interior of the country. Instability arose for two key reasons. First, gold reserves maintained by the U.S. Treasury fell to about $100 million from $190 million in 1890. At the time, the United States was on the gold standard, which meant that notes issued by the Treasury could be redeemed for a fixed amount of gold. The falling gold reserves raised concerns at home and abroad that the United States might be forced to suspend the convertibility of notes, which may have prompted depositors to withdraw bank notes and convert their wealth into gold. The second source of this instability was that economic activity slowed prior to the panic. The recession raised rates of defaults on loans, which weakened banks' balance sheets. Fearing for the safety of their deposits, men and women began to withdraw funds from banks. Fear spread and withdrawals accelerated, leading to widespread runs on banks.

M.G. BUCHOLTZ

In June 1893, bank runs swept through midwestern and western cities such as Chicago and Los Angeles. More than one hundred banks suspended operations. From mid-July to mid-August, the panic intensified, with 340 banks suspending operations. As these banks came under pressure, they withdrew funds that they kept on deposit in banks in New York City. Those banks soon felt strained. To satisfy withdrawal requests, money-center banks began selling assets. During the fire sale, asset prices plummeted, which threatened the solvency of the entire banking system. In early August, New York banks sought to save themselves by slowing the outflow of currency to the rest of the country. The result was that in the interior local banks were unable to meet currency demand, and many failed. Commerce and industry contracted. In many places, individuals, firms, and financial institutions began to use temporary expediencies, such as scrip or clearing-house certificates, to make payments when the banking system failed to function effectively.

From this description of events, it appears that September was perhaps the bottom of the crisis. September 1893 equates to 1893.75

    1893.75 + 3114 = 5007.75
    (5007.75/5175) * phi = 0.5980
    5175 – 5007.75 = 167.25
    167.25 * 0.5980 = 100.01
    5007.75 + 100.01 = 5107.76
    5107.76-3114 = 1993.76 = late September 1993.

Late 1993 witnessed the signing of NAFTA, the free trade agreement between Canada, Mexico, and the US. This agreement set in motion the events for what would turn into a crisis in 1994. Because of NAFTA, foreign investors were lending money to Mexico in exchange for US-dollar denominated bonds. A series of destabilizing political events in 1994 caused investors to demand a greater return on these bond purchases. The Mexican government acceded to these requests which caused an appreciation in the value of the peso. The stronger currency quickly led to a serious trade deficit and a flight of capital out of the country once investors realized the peso was overvalued. In late 1994,

39

the government abruptly devalued the peso which in turn posed risks to payments on US-dollar denominated debt. The US government ultimately provided Mexico with a $50 billion rescue package. The S&P 500 fell 1.5%, the first such drop since 1990. [4]

Taking the 1993.76 figure and passing it through Braden's algorithm, yields 2034.77 or **September 2034** as the date for another crisis event.

1907 saw another financial panic. As Moen and Tallman [5] of the Federal Reserve Bank of Atlanta describe:

The Panic of 1907 was the first worldwide financial crisis of the twentieth century. It transformed a recession into a contraction surpassed in severity only by the Great Depression.

In reading their synopsis of the crisis, it appears that November 1907 was the worst of the crisis. November 1907 equates to 1907.91.

$$1907.91 + 3114 = 5021.91$$
$$(5021.91/5175) * phi = 0.5997$$
$$5175 - 5021.91 = 153.09$$
$$153.09 * 0.5997 = 91.8$$
$$5021.91 + 91.8 = 5113.71$$
$$5113.71 - 3114 = 1999.71 = \text{late August 1999.}$$

At this 1907 timeframe, the seeds were sown for what would be an event in mid-1999. By mid-1999, the Nasdaq tech bubble was inflating in a dangerous fashion. It would just be a matter of time until it all fell apart.

In late December 1924, the Dow Jones Average ramped higher and surpassed its previous high made in late 1916. Using Braden's mathematical approach, this equates to an event in 2007. Late 2007 marked a peak in the equity markets and the start of what would be the sub-prime mortgage crisis. Just as 1924 signaled the start of a wild excursion higher on the market, 2007 signaled the start of a wild excursion lower on the market.

As another example, October 1957 marked the end of what had been 18 months of market weakness. October 1957 equates to 1957.83.

1957.83 + 3114 = 5071.83
(5071.83/5175) * phi = 0.6057
5175 − 5071.83 = 103.17
103.17 * 0.6057 = 62.49
5071.83 + 62.49 = 5134.32
5134.32-3114 = 2020.32 = March 2020.

March 2020 marked the end of a dramatic market swoon as COVID 19 made its appearance on the world stage.

Late December 1961 marked a peak on the Dow Jones after which it went on to decline by nearly 30%. December 1961 equates to 1961.95.

1961.95 + 3114 = 5075.95
(5075.95/5175) * phi = 0.6062
5175 − 5075.95 = 99.05
99.05 * 0.6062 = 60.04
5075.95 + 60.04 = 5135.99
5135.99-3114 = 2021.99 = the end of 2021.

Just like late December 1961 marked the start of a significant drawdown, the end of 2021 marked the start of a significant market drawdown. This calculation that points to late 2021 ties to the work of W.D. Gann and his 60-year cyclic advances. December 1961 plus 60 years equals December 2021. It is fascinating to see how two mathematical approaches can lead to the same conclusion.

December of 1965 marked an interim peak on the Dow Jones. August 1966 marked a swing low. Applying Braden's math to these dates reveals that the times around **August and November, 2023** could see inflection points on the equity market. Whether these will be small or large in scope is not known. Whether these events will be financial, geopolitical, or geological also remains unknown.

And, so it goes. Find a crisis event, use Braden's fractal math and one can arrive at the timing of a future crisis. In other words, just as the images in Figure 4-1 started from a basic shape, financial crises across time have all started from a fractal seed event in the annals of time.

I will leave you with one final example which deviates from Braden's technique of using a portion of the Equinox Precession as the starting pattern. Consider instead, a *Saros series* of eclipses. As the NASA Eclipse website notes:

Any two eclipses separated by one Saros cycle share very similar geometries. They occur at the same node with the Moon at nearly the same distance from Earth and at the same time of year.

A Saros series comprises a number of Saros cycles. As NASA describes:

A typical Saros series for a solar eclipse begins when new Moon occurs about 18° east of either the North or South Node. After ten or eleven Saros cycles (about 200 years), the first eclipse will occur near the south pole of Earth. Over the course of the next 950 years, each will be displaced northward by an average of about 300 km. The last eclipse of the series occurs near the north pole of Earth. The next approximately ten eclipses will be partial events with successively smaller magnitudes. Finally, the Saros series will end a dozen or more centuries after it began at the opposite pole. Due to the ellipticity of the orbits of Earth and the Moon, the exact duration and number of eclipses in a complete Saros is not constant. A series may last 1226 to 1550 years and is comprised of 69 to 87 eclipses. [6]

For example, Saros series 150 started on August 24, 1729 and will conclude at September 29, 2991 (1262 years). Taking the fractal seed event as being the start of the recession in January 1920 [7][8] and applying Braden's math, the calculation shows a future crisis event will occur in 100.17 years from January 1920. That equates to March 2020. It is humbling, to this author at least, to discover that two different time spans analyzed using Braden's methodology both point to March 2020 as the time for a significant event.

When I see fractal mathematics point to future events with a stunning degree of accuracy, I cannot help but think that the cosmos is the grand design of a higher power. Maybe that power is Allah. Maybe that power is God. Maybe that power is Vishnu. Call that power what you will. Moreover, our emotions are seemingly tied to cosmic intervals and events. The powerful money players in places like New York understand this and use it to their advantage.

Fractal mathematics suggests August and November 2023 will exhibit market trend turning points. Whether these inflection points will be financial, geopolitical, or geological remains unknown.

My cycles research has also led me to the work of Bruce Pugesek, an adjunct professor at Montana State University. [9] Professor Pugesek argues that significant turning points in economic history align to Fibonacci numbers and in some cases sums of two Fibonacci numbers. He further suggests that Lucas numbers also bear an alignment. He explains that people expect economic turning points to fall into an orderly arrangement. Fibonacci numbers and Lucas numbers are not orderly, unless one understands the technique for generating the numbers. Pugesek describes the alignment of these sequences to economic events as *irregular regularity*, a phenomenon which he says is at the core of fractal patterns.

The Fibonacci sequence of numbers was first contemplated 450-200 BC by Indian mathematicians seeking to analyze Sanskrit poetry syllables. In 1202, Italian mathematical thinker Fibonacci popularized this same sequence of numbers by way of his rabbit breeding example. He published his findings in a book entitled *Liber Abaci*.

The Fibonacci sequence of numbers is a recursive sequence in which a term of the sequence is the sum of the two prior terms. The numbers of this infinite sequence are:

0,1,1,2,3,5,8,13,21,34,55,89,144,233,377,610,987,1597...

In the mid-1800s, French mathematician and numbers theorist Edouard

Lucas introduced his interpretation of a recursive sequence in which a given term of the sequence is the sum of the two prior terms. The Lucas sequence is:

2,1,3,4,7,11,18,29,47,76,123,199,322,521,843,1364,2207,3571…

Pugesek's fractal method starts at a significant market event. He projects forward in time by cycles of a Fibonacci number F(n). He then projects backwards by smaller Fibonacci numbers. This forward and backward filling creates a fractal pattern of irregular regularity. The F(n) timeframe covered by one of these projections will equate to a combination of an F(n-1) and F(n-2) interval. Oftentimes, these cycle intervals will land on significant market events.

Consider the following example of the US equity market which uses the Fibonacci numbers F(n), F(n-1), and F(n-2), where F(n) is F(14) or 377.

If one takes the US equity market high point in the year 2000 and projects forward by three cycles of F(14) or 377 weeks, it can be seen that the first interval lands amidst the 2007 market highs. The second interval lands at a period of weakness in September 2014. The third interval is the late 2021 market high. Projecting a fourth cycle points to **March 15, 2029**.

From the starting point in 2000, apply two forward intervals of F(n-2) weeks, where F(n-2) equals 144 weeks. The end of the first interval lands right in the midst of the US invasion of Iraq, the second at a period of short term weakness in October, 2005.

From the end of this first 144-week interval, project two intervals of F(n-1) which is 233 weeks. The first interval lands at the terminus of the first 377-week interval. The second lands in December 2011.

From the terminus of the second 377-week interval, project *backwards* an interval of F(n-2) weeks, where F(n-2) equals 144 weeks. This backwards projection lands right at the 2009 market lows.

From the terminus of the second 377-week interval project two intervals of F(n-2) weeks, where F(n-2) equals 144 weeks. The second of these 144 week periods lands at the March 2020 COVID panic lows.

From the terminus of second 377-week interval, project forward one interval of F(n-1) weeks, where F(n-1) equals 233 weeks. This lands in March 2019. A projection of 144 weeks forward lands on the terminus of the third 377-week cycle which was the 2021 market high.

From this late 2021 terminus point, projecting forward 377 weeks lands in **March 15, 2029.**

From the March 15, 2029 terminus, project *backwards* one interval equal to F(n-1) weeks which is the Fibonacci number 233. This will land in **late September, 2024**.

From the March 15, 2029 terminus, project backwards two intervals of F(n-2) weeks. The first projection of 144 weeks takes the time to around **June 21, 2026**. The second projection lands in **mid-September of 2023**. This is very close to the date intervals identified earlier in this chapter using Braden's methodology. From the late 2021 F(14) weekly interval, if one considers that the F(n-2) interval is actually the sum of the F(n-3) and F(n-4) intervals, an attempt can be made to postulate when the current market mis-behavior might end. Late December 2021 plus 55 weeks (which is F(n-3)) lands in early **January 2023**. This date ties to the work of Professor Weston, whom you will read about in Chapter 9.

Pugesek's work is fascinating, in that it also delves deep into world history. In particular, historians regard Rome to have been founded in August of 754 BC. Western Rome is regarded by historians to have fallen in 411 AD when sacked by the Visigoths. This total timespan is 1165 years which is a fractal value arrived at by multiplying the 13th Fibonacci number by 5 (233 x 5 = 1165). The Eastern Roman empire survived until 1453 AD. The time span from the founding of Rome to the year 1453 AD is 2207 years which is the 16th Lucas sequence term. The year 1453 was also the year when the Kuwae volcano (near the south Pacific islands of Epi and Tongoa) erupted which led to global climate disruption.

Geologists estimate that the Taupo volcano (situated on the north island of New Zealand) erupted in 180 AD (+/- 4 years). Pugesek applies Fibonacci sequence terms to a date of 184 AD. He notes that if one adds five intervals of 233 years (233 is the 13th Fibonacci term) to 184 AD, the result computes to 1349 AD which marked the peak of severity of the Black Death plague in Europe. Pugesek extends his argument by noting that after the plague decimated Europe, a new nation (United States) formally began self-government in 1781 when British forces surrendered. From 1349 to 1781 is 432 years (3 x 144 where 144 is the 12th Fibonacci term). Pugesek notes that intervals of 233 years (13th Fibonacci term) are important in history. From 1349 AD, the next 233 year interval was 1582. This notable year marked the introduction of the Gregorian calendar by Pope Gregory. The next 233 year interval was 1815. This monumental year marked the defeat of Napoleon at the Battle of Waterloo. This year also marked the eruption of the Tambora volcano in Indonesia. The next 233 interval will take us to the year 2048. What will happen then?

Using the Fibonacci and Lucas numbers, one can further explore intervals of history. For example, October 1929 recorded a market peak now known as Black Thursday. It would take until 1932 for a significant bottom to appear. One can apply Pugasek's method of forward and backward projections as follows:

From the market lows in 1932, project forward F(10), which is 55 years. This takes one to 1987 when the markets swooned in an event now remembered as the *Crash of '87*. Project backwards by F(9) or 34 years and one arrives at 1953 and a market inflection point that came with the ending of the Korean War. Project ahead by an interval of F(8), or 21 years to arrive at 1974 which was a monumental year politically and economically. From the 1987 terminus of the F(10) projection, add F(9) or 34 years and one lands in 2021, which as we now know marked a turning point on the market along with the Ukraine invasion. Now back up by F(8) or 21 years and the 2000 tech bubble collapse comes into focus. The connection between Fibonacci/Lucas recursive number patterns is riveting. The world is anything but random.

I will conclude this chapter by further observing that Fibonacci and Lucas terms do not always have to be taken as years or weeks of time. They could be interpreted as days or hourly chart bars of time. For example, the heliocentric orbital period of Mercury is 88 days, which is one day short of the 11th Fibonacci term. The heliocentric orbital period of Venus is 225 days which is very close to 3 x 76 (the 9th Lucas term). Mars orbits the Sun in 687 days which closely aligns to 9 x 76 days, where 76 is the 9th Lucas term. When applying Lucas and Fibonacci numbers to occurrences in the cosmos, we are limited only by the extent of our imaginations.

I will conclude with a practical image from Pugesek's paper which illustrates the forwards and backwards interval applications.

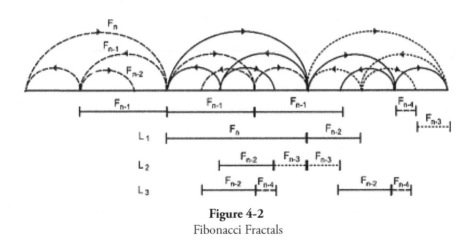

Figure 4-2
Fibonacci Fractals

CHAPTER FIVE
Trend Changes

In previous editions of this Almanac, I have included a paragraph advising readers that planetary cyclical events should be acted on if a price trend change is also visible on the stock, commodity future, or index being looked at. Discussions with people who have purchased previous Almanacs and with subscribers to my newsletters have revealed that the subject of trend change needs more focus in my written offerings. This chapter focuses on price trend and illustrates how technical chart indicators can be used to help determine when the price trend changes.

Financial market data platforms come programmed with a variety of *moving averages* ranging from *simple* to *exponential* to *smoothed*.

On a given trading day the simple moving average is the mathematical average of the past 'n' price bars on a price chart of a stock, commodity, or futures contract. The next day, the calculations discard the first value in the data and add the most recent data point. There is considerable flexibility in determining how many data points to use in the average calculation. An 18-period moving average will provide for more frequent moves of price above or below the average. A 34-period moving average is less sensitive, but still a reliable time frame.

An exponential moving average is similar but the mathematical calculation underlying the average places greater weight (emphasis) on the more recent of the 'n' data points.

A smoothed moving average is an average of an average. Suppose an 18 day moving average has been calculated where n=18 price bars. After a longer period of time has elapsed, suppose an 18 day moving average of the 18-day moving average data points is calculated. This smoothed moving average reduces the variability of the data points and helps traders better identify trend directions.

Traders will often regard the trend to have turned positive (favorable) when price moves above the moving average. As price retreats below a moving average, traders will regard the trend to have turned negative.

Market data platforms also come programmed with a variety of *stochastic* and *oscillator* functions. Stochastic and oscillator functions compute the price range of data over a period of time. The price at a given day is then expressed as a percentage of this data range.

What follows in this brief chapter is an examination of price trend changes on some selected stocks in the context of planetary events. It will take you some time to fully explore the assortment of technical chart indicators available on your particular data platform. Once you have found ones you are comfortable using, as a planetary event draws near you will be able to assess whether or not the trend is changing based on what the chart indicators are showing.

Figure 5-1 presents daily price action on Devon Energy (NYSE:DVN) from March through October 2022. The rectangular shapes overlaid on the chart denote Mercury retrograde events. The price moved higher during the retrograde event in May 2022, but within days of retrograde completing, the price of Devon Energy shares started to decline sharply.

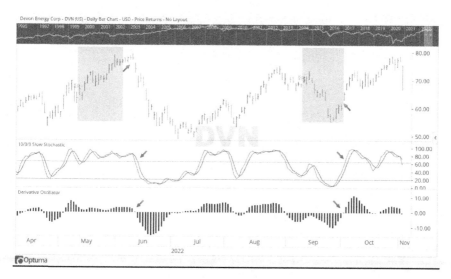

Figure 5-1
Devon Energy (NYSE:DVN)

The panels beneath the price chart illustrate the *Slow Stochastic* and the *Derivative Oscillator*. The decline in price following the end of retrograde was signaled by the Slow Stochastic crossing a decision-making threshold value (I set this threshold value to 65 on my charting). At the same time, the Derivative Oscillator bars shifted from being above the zero line to being beneath the zero line. At the right side of the chart, it is evident that the price trend changed from positive to negative prior to the onset of another Mercury retrograde. Price reached a low point during the retrograde event. As price started to gain traction, the Slow Stochastic and Derivative Oscillator both signaled a buying opportunity.

Figure 5-2 presents daily price action on semiconductor maker Micron Technology (NYSE:MU). The rectangular shape overlaid on the chart denotes the Venus retrograde event that started in late 2021. Initially, price continued trending higher as the retrograde event manifested. But, in early January, 2022 the Slow Stochastic crossed the threshold value of 65 signaling that the trend was changing. The Derivative Oscillator confirmed this trend change which saw price drop nearly $20 per share. The retrograde event concluded on January 29, 2022.

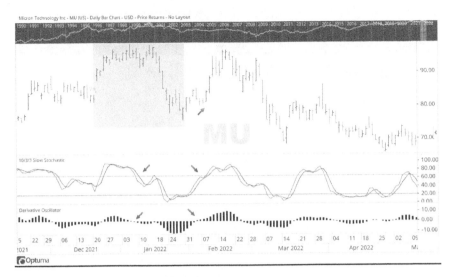

Figure 5-2
Micron Technology (NYSE:MU)

On February 7, the Slow Stochastic and Derivative Oscillator both signaled a buying opportunity. A trader or investor heeding the Venus retrograde event would have sidestepped a share price decline and then re-entered the trade at a lower price.

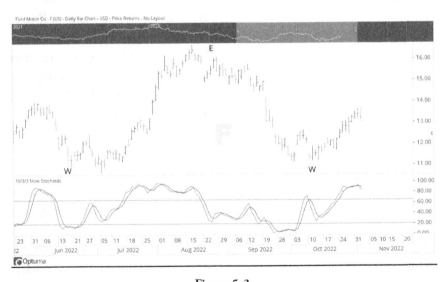

Figure 5-3
Ford Motor Company (NYSE:F)

Figure 5-3 presents daily price action of shares in the Ford Motor Company (NYSE:F). The chart has been overlaid with notations 'W' and 'E' which refer to events of Mercury at its greatest easterly and westerly elongation. Times of Mercury elongation maxima can often lead to trend changes. In June 2022, Mercury reached its greatest westerly elongation (W). A trader buying the shares at this event would have experienced an increase in share price. But, price quickly fell back on itself. Note that despite the appearance of this Mercury westerly elongation event, the Slow Stochastic never did get above the decision-making threshold value of 65. Heeding this Slow Stochastic action would have kept an alert trader out of the stock. In August, 2022 Mercury recorded its greatest easterly elongation (E). The actual trend change was signaled several days prior as the Slow Stochastic crossed beneath the threshold value of 65. The stock price remained weak until October. An alert trader noting the arrival of the easterly elongation event, would have exited the stock. In October, 2022 Mercury recorded its greatest westerly elongation (W). At this time of writing, price seems to be bottoming. However, the Slow Stochastic is still beneath the threshold of 65. A buy signal will materialize only when the Stochastic gets above the 65 threshold value.

Technical chart analysis is a powerful tool that can keep a trader or investor out of trouble. Not every planetary event will immediately align to a buying opportunity. By watching the chart indicators, traders and investors can embrace price trend changes and make sound decisions as planetary events occur.

CHAPTER SIX
Venus Cycles

Cycles of Venus play a key role in forecasting trend changes on financial markets.

Venus orbits the Sun in 225 days relative to an observer standing at a fixed venue like the Sun. This is its *heliocentric* orbital time.

To an observer situated on Earth (a moving frame of reference), Venus appears to take 584 days to orbit the Sun. This is its *geocentric* orbital time.

During this 584-day geocentric orbital period, there will be periods of time when Venus is situated between Earth and Sun. This is its *Inferior Conjunction*. As Venus slowly moves out of this conjunction, it will become visible as the *Morning Star*.

During its 584 day geocentric orbital period, there will be periods of time when the Sun is situated between Venus and Earth. This is its *Superior Conjunction*. As it moves out of this conjunction, it becomes visible as the *Evening Star*.

Human psyche is influenced by Venus and its shifts from being visible in the morning to being visible in the evening and vice versa.

When one considers the pattern of Superior and Inferior conjunctions over time, the influence of Venus becomes even more intriguing. Multiplying the 584-day geocentric cycle by 5 yields a value of 2920. Dividing 2920 by the 365-day orbital period of Earth yields a value of 8. The ratio of the Venus geocentric orbital period to the orbital period of Earth is thus a 5:8 ratio. Multiplying eight by five yields 40, a number cited frequently in Biblical texts.

Venus was at Superior Conjunction on March 28, 2013 (8 Aries), October 25, 2014 (1 Scorpio), June 6, 2016 (16 Gemini), January 8, 2018 (18 Capricorn), August 14, 2019 (27 Leo), March 26, 2021 (6 Aries), and October 22, 2022 (4 Libra).

Venus was at Inferior Conjunction on June 6, 2012 (15 Gemini), January 11, 2014 (21 Capricorn), August 15, 2015 (22 Leo), March 25, 2017 (4 Aries), October 26, 2018 (3 Scorpio), June 4, 2020 (13 Gemini), and January 9, 2022 (18 Capricorn).

If one plots these groups of Superior Conjunction events on a zodiac wheel, it becomes evident that they can be joined to form a 5-pointed star called a pentagram. Likewise, the Inferior Conjunction events can also be plotted and joined to form a pentagram. Such are the mysteries of our cosmos.

As Venus orbits around the Sun following the ecliptic plane, it moves above and below the plane. The high points and low points made during this travel are termed *declination maxima* and *minima*. Declination maxima and minima play key roles in price trend changes on financial markets.

    ✿ In early 2018, a Venus Superior Conjunction and declination minimum was followed closely by a steep 300 point sell-off on the S&P 500

☼ A significant change of trend immediately preceded an Inferior Conjunction and declination minimum in October 2018. This was followed by an acceleration of trend to the downside

☼ In July 2019, talk of a recession and an impasse with China over trade disputes caused a sharp sell-off on the S&P 500. This sell-off event overlapped with a Venus Superior Conjunction and a Venus declination maximum

☼ A Venus Superior Conjunction in early December 2019 failed to deliver a meaningful reaction. The Federal Reserve was busy ramping up liquidity to the banking system which had the effect of dampening the influence of planetary cycles on the markets

☼ In early May 2020, Venus made a declination maximum and the S&P 500 exhibited a short, sharp 200-point drop. The damage could have been worse in the absence of the Federal Reserve's liquidity injection

☼ The Inferior Conjunction event on June 3, 2020 delivered a 200 point drop to the S&P 500. But again, the Federal Reserve came to the rescue with more fiat liquidity

☼ The days immediately following the January 14, 2021 Venus declination low were met with the S&P 500 dropping just 200 points (high to low)

☼ The Superior Conjunction event of March 26, 2021 was immediately preceded by the Dow Jones dropping 1100 points (high to low) over six trading sessions

☼ The Venus declination maximum on June 4, 2021 was immediately followed by a 1500-point (high to low) drawdown on the Dow Jones Average

☼ The Venus declination minimum in early November, 2021 was followed by a 200 point drawdown on the S&P 500. A brief recovery attempt followed that lasted into early January 2022 but failed to hold

☼ The Venus Inferior Conjunction on January 8, 2022 confirmed that the trend had indeed changed on equity markets

☼ The Venus declination maximum event on July 23 marked the mid-point of a counter-trend rally that saw the Dow Jones rally over 4000 points

☼ The Venus Superior Conjunction event on October 22, 2022 yielded chaos in the UK as newly minted Prime Minister Lizz Truss resigned. The S&P 500 was signaling that a turning point was coming. Five days later, on October 27, a counter-trend rally on the S&P 500 failed and the market softened.

Venus will be at its declination minimum again on December 13, 2022.

Another cyclical event pertaining to Venus is its retrograde events. From our vantage point on Earth, we describe the position of Venus relative to one of the twelve star constellations in the sky. There will be one (occasionally two times) time per year when Earth and Venus pass by each other. As Venus starts to lap past Earth, owing to the different orbital speeds of Earth and Venus, there will be a period of time when we see Venus in what appears to be the previous constellation to where it was just prior to starting to lap by Earth. For example, we might start off seeing Venus against the star constellation of Gemini. As Venus begins to lap past Earth, we will see Venus against the star constellation of Taurus. As Venus passes by Earth, we will see Venus again in Taurus. Of course, Venus has not physically reversed course and moved backwards. This is an optical illusion created by the different orbital speeds of Venus and Earth.

These brief illusory periods of backwards motion are what astrologers call *retrograde* events. To ancient societies, retrograde events were of great significance as human emotion was often seen to be changeable at these events.

There is a curiously strong correlation between equity markets and Venus retrograde. Sometimes Venus retrograde events encompass a sharp market inflection point. Sometimes a market peak or bottom will follow closely behind a retrograde event. Sometimes a peak or bottom will immediately precede a retrograde event. When you know a Venus retrograde event is approaching, use a suitable chart technical indicator to determine if the price trend is changing.

Figure 6-1 illustrates price behavior of the E-mini S&P 500 Index in early 2022 as Venus was retrograde. This chart has been fitted with the Slow Stochastic indicator. Note that mid-way through the Venus retrograde event the Slow Stochastic signaled a trend change.

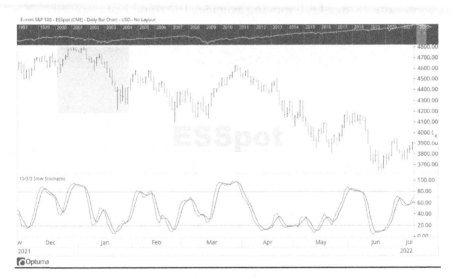

Figure 6-1
E-mini S&P 500 Index and Venus retrograde

Knowing that the potential exists for sizeable moves, aggressive traders can avail themselves of these retrograde correlations. Less aggressive investors may simply wish to place a stop loss order under their positions to guard against sharp price pullbacks.

The influence of Venus also extends beyond just the equity markets. Commodities can also be influenced:

☼ In early 2022 just as the Venus retrograde event started, Gold prices started to trend higher, gaining over $200 per ounce

☼ As the Venus retrograde event started, Oil prices started to trend higher, moving from $67 per barrel to $130 per barrel

☼ As the Venus retrograde event in early 2022 wrapped up, Wheat prices started to rally, moving from $7 per bushel to $13 per bushel. Figure 6-2 shows the price performance of Wheat futures. Just as retrograde was done, the MAC-D crossed positive and above the zero line. Prices moved above the 28-day Hull moving average. The trend had changed.

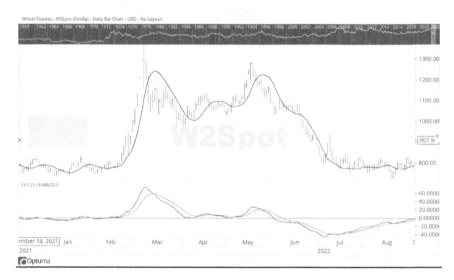

Figure 6-2
Venus retrograde and Wheat Futures

For 2023:

☼ **Venus will be retrograde from July 23, 2023 through September 3, 2023**

☼ **Venus will be at Inferior Conjunction starting August 13, 2023**

☼ **The next Superior Conjunction will be early June 2024**

☼ **Venus will be at its maximum declination between April 24 and May 24. The exact maximum will occur at May 10 with Venus being 26 degrees above the ecliptic plane.**

CHAPTER SEVEN
Mercury Cycles

Mercury is the smallest planet in our solar system. It is also the closest planet to the Sun. As a result of its proximity to the powerful gravitational pull of the Sun, Mercury moves very quickly, completing one heliocentric cycle of the Sun in 88 days.

Mercury has an eccentric orbit in which its distance from the Sun will range from 46 million kms to 70 million kms. When Mercury is nearer to the Sun (46 million kms away), it is moving at its fastest (56.6 kms per second). When Mercury is farther from the Sun (70 million kms away), it is moving slower (38.7 kms per second). [1]

Mercury at its closest orbital point to the Sun is called *Mercury Perihelion*. Mercury at its farthest orbital point to the Sun is called *Mercury Aphelion*.

In 2020, Mercury was at Perihelion just as the equity markets were reaching a peak ahead of the COVID panic selloff. The selloff lows came just as Mercury was nearing Aphelion. Mercury Perihelion and Aphelion events do not all seamlessly correlate to all market inflection points. Nevertheless, there is enough past data to suggest these dates should be anticipated by traders and investors.

For 2022, Mercury was at Perihelion at: January 16, April 14, July 11, and October 7. Aphelion dates were: March 1, May 28, August 24, and November 20.

The January 16 Perihelion event aligned with the start of a significant decline on the S&P 500. The April 14 Perihelion event aligned to a brief counter-trend rally attempt. Three days after the July 11 Perihelion event, the S&P 500 embarked on a 600 point counter-trend rally. The October 7 Perihelion event triggered a decline that sent the S&P 500 to fresh lows for 2022.

The March 1 Aphelion event marked the start of a renewed dip lower on the S&P 500. The May 28 Aphelion event market another turning point and the end of a countertrend rally. Two days after the August 24 Aphelion event, what appeared to be perhaps the start of a rally failed miserably.

For 2023:

> ☼ **Mercury will be at Perihelion: January 3, April 4, June 28, Sept 24, and December 21**

> ☼ **Mercury will be at Aphelion: February 16, May 15, August 11, and November 7.**

Related to Mercury's orbit is its *elongation*. As discussed earlier in Chapter 1, elongation refers to the angle between a planet and the Sun, using Earth as a reference point. Figure 7-1 illustrates the concept of elongation. Times when Mercury is at its greatest elongations bear a strong correlation to short term turning points on financial markets.

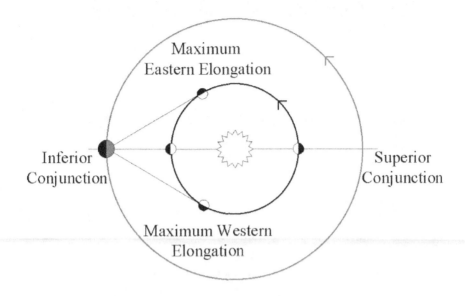

Figure 7-1
Elongation of Mercury

In 2022:

- ✿ Mercury was at its greatest easterly elongation January 7, April 29, August 27, and December 21
- ✿ Mercury was its greatest westerly elongation February 17, June 16, and October 9.

Figure 7-2 illustrates some of these 2022 events overlaid on a chart of the E-mini S&P 500.

The peak of the market in January 2022 occurred several days prior to a greatest easterly elongation event (E). A countertrend rally failed in February as a greatest westerly event (W) loomed. A greatest easterly elongation event (E) in late April added intensity to an existing downtrend. A greatest westerly event (W) in mid-June marked the start of a brief countertrend rally. The greatest westerly event (W) in early October set the stage for a 300 point rally on the S&P 500.

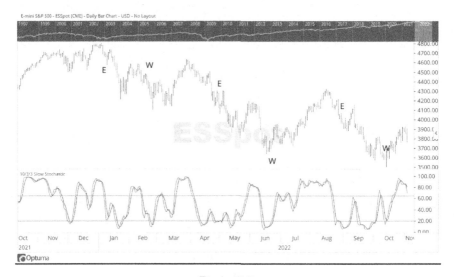

Figure 7-2
2022 Mercury Elongation events

☼ **For 2023, Mercury will be at its greatest easterly elongation April 11, August 10, and December 4.**

☼ **For 2023, Mercury will be at its greatest westerly elongation January 30, May 29, and September 22.**

In addition to elongation, events there will be retrograde events. Mercury retrograde has been popularized by classical astrologers who tell clients not to sign important contracts during Mercury retrograde, not to not cross the street, to not leave their houses and so on. While I tend to ignore this mundane talk, I have noticed a striking correlation between financial market behavior and Mercury retrograde events.

There will be three (occasionally four) times during a year when Earth and Mercury pass by each other. As Mercury starts to lap past Earth, owing to the different orbital speeds of Earth and Mercury, there will be a period of time when we see Mercury in what appears to be the previous constellation. For example, just as Mercury starts to lap past Earth, we might start off seeing Mercury against the star constellation of Gemini. As Mercury begins to lap past Earth, we will see Mercury against the star constellation of Taurus. As Mercury passes by Earth, we will see Mercury again in Taurus. Of course, Mercury has not physically reversed course and moved backwards. This is an optical illusion created by the different orbital speeds of Mercury and Earth.

These brief illusory periods are what astrologers call *retrograde* events. To ancient societies, retrograde events were of great significance as human emotion was often seen to be changeable at these events.

Retrograde events involving Mercury very often align to short term price trend changes developing. Sometimes Mercury retrograde events encompass a sharp market inflection point. Sometimes a market peak or bottom will follow closely behind a retrograde event; sometimes a peak or bottom will immediately precede a retrograde event. Mercury retrograde events can be highly unpredictable as to when the impact on a financial instrument will appear. Is it possible that our DNA is hard-wired such that we feel uncomfortable at Mercury retrograde events? Does this emotional discomfort compel us to buy or perhaps sell on the financial markets?

Knowing that a Mercury retrograde event is approaching, traders and investors should remain alert for indications of a trend change using chart technical indicators.

Figure 7-3 illustrates three Mercury retrograde events from 2021 and 2022 overlaid on a chart of the E-mini S&P 500. The chart has been fitted with the Slow Stochastic indicator. By the time Mercury turned retrograde in January 2022, the trend had already turned negative. A buy signal materialized late in the retrograde event. As retrograde completed, the trend again turned negative. The trend was to the downside as the May retrograde event arrived. In the midst of the retrograde event, a buy signal materialized. But, in classic Mercury behavior, no sooner had retrograde completed than the trend turned decidedly negative again. In September, the market looked to be gathering some tentative strength, but Mercury being retrograde quickly caused a resumption of the negative trend. No surprise then that classical astrologers often refer to Mercury as *the trickster*.

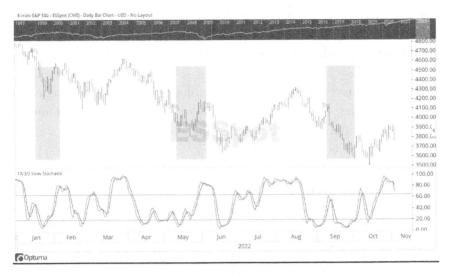

Figure 7-3
E-mini S&P 500 and Mercury retrograde

Mercury retrograde events can sometimes be seen influencing commodity markets. For example, Figure 7-4 illustrates a Copper futures price chart overlaid with Mercury retrograde events from 2022. Of the three retrograde events in 2022, only the event in May proved tradeable. Mid-way through the retrograde event, the Slow Stochastic delivered a buy signal. Investors following copper-related mining stocks could have used this retrograde event for some short term trading. Note how the price of Copper began to tumble just as the retrograde event concluded. Such is the nature of Mercury retrograde.

Figure 7-5 illustrates the correlation between Mercury retrograde and Wheat futures prices. Figure 7-5 has been fitted with the Slow Stochastic indicator. Taking a disciplined approach of buying when the Stochastic crossed up and through the 65 threshold value would have presented a trader with some volatile, but interesting, opportunities.

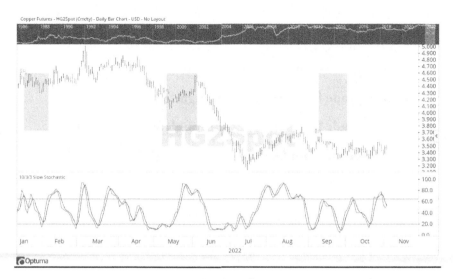

Figure 7-4
Mercury retrograde and Copper futures

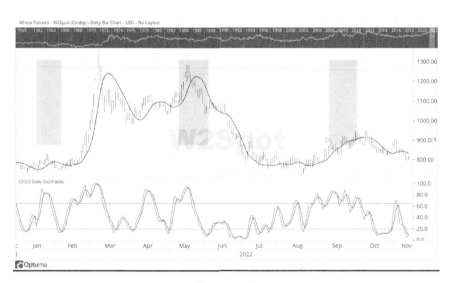

Figure 7-5
Mercury retrograde and Wheat Futures

For 2023, Mercury will be:

- ☼ retrograde from December 29, 2022 through February 17
- ☼ retrograde from April 21 through May 14
- ☼ retrograde from August 23 through September 14
- ☼ retrograde from December 13 through January 1, 2024.

CHAPTER EIGHT
The Pleiades

The Pleiades is a cluster of stars situated in the constellation of Taurus. On the geocentric zodiac wheel, the position of 27 degrees Taurus aligns with the Pleiades star cluster.

The Pleiades are sometimes called the Seven Sisters. Many cultures from around the world have mythological tales that incorporate the Pleiades. For example, the Cree indigenous people of western Canada regard the Pleiades as a hole in the sky from which they came. As the story goes, Sky Woman spotted Earth and expressed a desire to visit the planet. With help from Spider Woman, who spun a web, Sky Woman was able to complete her journey to planet Earth.

The chart in Figure 8-1 depicts daily price action on Gold futures. The chart has been overlaid with two events of Mars passing the Pleiades at 27 Taurus; one event in early 2021 and the other in mid-2022. In March 2021, Gold prices were trending down. The downtrend was arrested shortly after Mars passed the Pleiades. With Mars still within 5-degrees of exact conjunction to the Pleiades, Gold recorded a turn of fortune and went on to rally over $200 per ounce. In mid-2022, a Mars-Pleiades conjunction triggered a sharp drawdown in Gold prices.

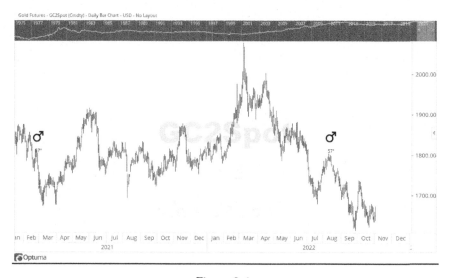

Figure 8-1
Gold and Mars passing the Pleiades

The chart in Figure 8-2 illustrates price action on the E-mini S&P 500. Note how Mars passing the Pleiades aligns to the August 2022 highs.

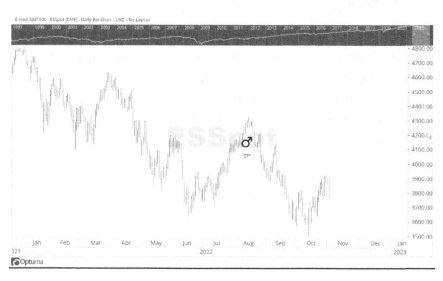

Figure 8-2
E-mini S&P and Mars passing the Pleiades

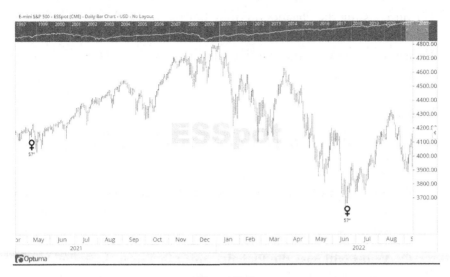

Figure 8-3
E-mini S&P and Venus passing the Pleiades

As Figure 8-3 illustrates, May 2021 weakness aligned to Venus passing the Pleiades. The June 2022 lows on the S&P 500 also align to Venus passing the Pleiades.

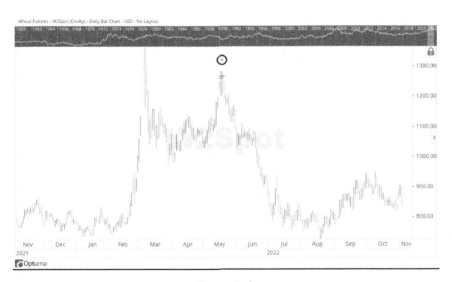

Figure 8-4
Wheat futures and Sun passing the Pleiades

73

The Sun passing the Pleiades point in Taurus also can bear an alignment to commodity price action. Figure 8-4 illustrates Wheat futures prices. Notice how Sun conjunct the Pleiades in May, 2022 aligns to a trend change.

In 2023:

☼ **Sun will pass the Pleiades point between May 14 and May 25**

☼ **Sun will be 90-degrees square to the Pleiades from February 12 to 20. Sun will again be square the Pleiades from August 16 to 24**

☼ **Venus will pass the Pleiades point in Taurus between April 4 and April 13. Venus will be 90-degrees square the Pleiades from January 21 to 28. Venus will again be square from July 6 to 22**

☼ **Mars will not pass the Pleiades point in 2023.**

CHAPTER NINE
Professor Weston's Cycles

In 1921, a mysterious person from Washington, D.C. using the name Professor Weston wrote a paper in which he analyzed many decades worth of past price data from the Dow Jones Average. Who exactly Weston was, I will likely never know; another one of those figures who emerged to write his ideas down before vanishing into the ether.

His work was based on the premise that the Dow Jones price data was comprised of a series of interwoven, overlapping cycles. He applied cosine Fourier mathematics to the data to delineate the cycles. His analysis identified a 10-month, 14-month, 20-month and 28-month cycle pattern. Using the OPTUMA software, I have been able to demonstrate that these 10,14,20,and 28-month cycle intervals are indeed still evident on a monthly chart of the Dow Jones Average.

As an aside, it is curiously interesting that 11 and 29 are Lucas sequence components. Fibonacci components include the values 13 and 21. The subtle connections that weave their way through Nature are more than intriguing.

Perhaps Weston knew W.D. Gann personally. Perhaps he just knew of him. In any case, Weston followed the 20-year Gann Master Cycle of Jupiter and Saturn.

Weston further broke this long cycle into two components of 10 years.

He believed investors can expect:

✿ a 20-month market cycle to begin in November of the 1st year of the 10-year cycle

✿ another 20-month cycle to begin in November of the 5th year of the 10-year cycle

✿ 28-month cycles to begin in July of the 3rd and 7th years of the 10-year cycle

✿ a 10-month cycle to begin in November of the 9th year of the 10- year cycle

✿ a 14-month cycle to begin in September of the 10th year of the 10-year cycle.

To put this into perspective, a new Gann Master Cycle began on November 1, 2020 as heliocentric Jupiter and Saturn made a 0-degree aspect. The entire cycle will run until November 2040.

Following Weston's methodology for the first half of the overall Master Cycle:

✿ the first 20-month cycle will start in November 2020 and go to until July 2022

✿ a 28-month cycle will run from July 2022 through November 2024

✿ a 20-month cycle will then follow until July 2026

✿ a 28-month cycle will run July 2026 through November 2028

☼ a 10-month cycle will then run through until September 2029

☼ lastly, a 14-month cycle will last until November 2030.

Looking at a chart of the S&P 500, one can delineate a cycle than runs from the lows of November 2020 to the lows of late June 2022. We are now in the 28-month cycle that will terminate at or near November 2024.

Weston also postulated that in the various years of a 10-year segment of the overall Master Cycle, there would be market maxima as listed in Figure 9-1.

YEAR OF CYCLE	MAXIMA	MAXIMA
1	March	October
2		May
3	January	September
4	April	November
5	May	November
6		June
7	January	September
8		June
9	April	
10	February	August

Figure 9-1
Weston's Secondary Cycles

Weston calculated these events using cycles of Venus. He argued that the 16th harmonic of a 10-year period (120 months) was actually the heliocentric time it takes for Venus to orbit the Sun (120 x 30 / 16 = 225 days). Back-testing has shown that these dates should be taken with a generous time span of perhaps +/- 3 weeks. In other words, an October high might manifest in late September or perhaps in early-November.

Within the new Master Cycle that began in November 2020, Weston's work cautions investors to be alert for market maxima in March 2021 and in October 2021.

The S&P 500 recorded an interim high in late February, 2021. By early March, the S&P had fallen over 200 points. A swing low was realized on March 4, 2021. An interim high maxima point was seen in September 2021, but the end of September recorded an interim low.

This apparent confusion could very well be related to the Federal Reserve pumping massive fiat liquidity into the system. It is also possible that Weston's predictions might be experiencing an inversion. That is, his dates might be delivering price minima instead of maxima.

Weston's work next suggests a market maximum is due (possibly could be a minimum) in May 2022. In fact, what materialized in May was a low after the S&P had fallen 500 points.

January and September 2023 are forecast to again deliver market maxima (perhaps they could be minima events) points as per Weston's work.

CHAPTER TEN
Synodic Cycles and Bradley

As discussed earlier, planetary movements can be examined in terms of synodic cycles (geocentric) and sidereal cycles (heliocentric).

The time it takes from Venus recording a conjunction with Sun until that same conjunction occurs again is 584 days. Mars takes 780 days from a Sun/Mars conjunction to the next Sun/Mars conjunction. Sun/Saturn conjunctions are 376 days apart. Other outer planets take 367 to 399 days.

Synodic cycles involving celestial bodies such as Sun, Mercury, and Venus relative to slower moving bodies such as Pluto, Saturn, and Jupiter are of particular interest in identifying trend turning points on equity and commodity markets. A trailblazer in correlating this synodic behavior to the markets was the late American financial astrologer, Jeanne Long. [1]

Figure 10-1 illustrates Oil prices and the Sun-Pluto synodic cycle. The aspects visible on this chart are those of 0, 60, 90, and 120-degrees. All aspects have been drawn to within orb (+/-5 degrees of exact alignment).

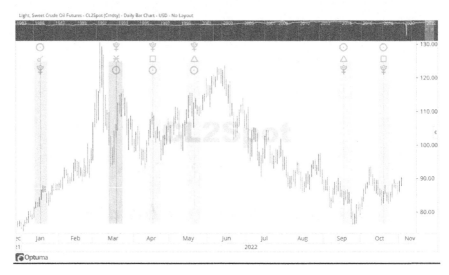

Figure 10-1
Sun-Pluto aspects and Oil prices

In January 2022, Oil prices were trending higher. The Sun-Pluto conjunction had little effect on the strongly trending market. In March 2022, Oil prices peaked. Mere days later, Sun and Pluto formed a 60-degree aspect. Price rebounded slightly, but then promptly failed as the 60-degree aspect concluded. The 90-degree square aspect in April produced a brief run-up in price which quickly reversed itself. The 120-degree trine aspect in May did not disturb the uptrend that was in place at the time. A low swing point in September came as a 120-degree trine aspect concluded. At this time of writing, a square aspect is about to manifest, followed by a 60-degree aspect in November.

Figure 10-2 illustrates Silver prices in the context of aspects between Venus and Jupiter. A 60-degree aspect in early February, 2022 aligned to a swing low. Silver prices went on to record a double top formation, with the second top coming in April. Several days after this second top, Venus reached a 0-degree aspect with Jupiter and price swooned. A square aspect between these two planets in July prompted a brief rally, which then faded at a 120-degree trine aspect in August.

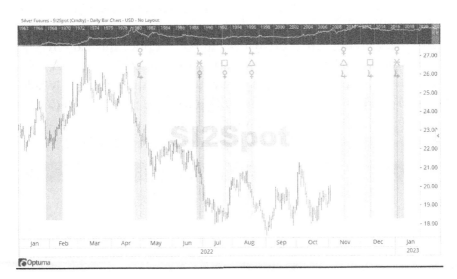

Figure 10-2
Venus-Jupiter aspects and Silver prices

Figure 10-3 illustrates aspects between Sun and Jupiter from late 2021 through June 2022 as overlaid on a chart of E-mini S&P 500 futures. The connection is readily apparent.

Figure 10-3
Sun-Jupiter aspects and E-mini S&P 500

Figure 10-4 illustrates aspects between Mercury and Saturn during 2022 overlaid on a chart of Soybean futures. A price spike and reversal in February 2022 aligns to a Mercury-Saturn 0-degree aspect. A swing low in April occurred at a 60-degree aspect. The subsequent rally faded at a 90-degree square event. The 2022 price high on Soybeans came in June just as a 60-degree aspect was concluding.

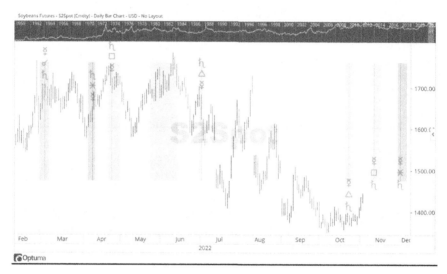

Figure 10-4
Mercury-Saturn aspects and Soybeans

Over and above Jeanne Long's excellent work from years ago, my back-testing has shown that:

- ✿ heliocentric Jupiter-Neptune aspects influence trend changes on Wheat and Corn

- ✿ Mercury-Saturn aspects influence short term trend changes on Euro, Pound, Australian Dollar, Canadian Dollar currency futures, Corn futures, and meats futures (Live Cattle, Feeder Cattle, Lean Hogs)

- ✿ Mercury-Jupiter aspects influence Wheat futures, 10-Year Treasury Notes, and 30-Year Bonds

✿ Sun-Neptune aspects influence Gold futures and Coffee futures

✿ Sun-Pluto aspects influence Copper, Cocoa, and Sugar futures

✿ Sun-Jupiter aspects influence the Nasdaq.

For 2023, the above described aspects will occur on the following dates:

PLANETARY PAIR	ASPECT	DATES
Sun-Pluto	0 degrees	January 13-23
	60 degrees	March 16-23
	60 degrees	November 16-25
	90 degrees	April 15-25
	120 degrees	May 16-26
	120 degrees	September 16-26
	180 degrees	July 16-26
Sun-Jupiter	60 degrees	January 20-28
	60 degrees	June 26-July 5
	0 degrees	March 31-April 22
	90 degrees	July 30-August 12
	120 degrees	September 3-13
	120 degrees	December 21-Jan 3, 2024
	180 degrees	October 28-November 8
Sun-Neptune	60 degrees	January 9-16
	60 degrees	May 13-21
	0 degrees	March 10-20
	90 degrees	June 13-23
	90 degrees	December 11-22
	120 degrees	July 15-25
	120 degrees	November 12-22
	180 degrees	September 14-24

PLANETARY PAIR	ASPECT	DATES
Venus-Jupiter	0 degrees	February 22-March 9
	60 degrees	January 2-6
	60 degrees	May 1-7
	60 degrees	August 12-September 25
	90 degrees	June 4-18
	120 degrees	October 15-27
	180 degrees	December 3-14
Mercury-Saturn	0 degrees	February 24-March 6
	60 degrees	April 3-7
	60 degrees	May 7-24
	60 degrees	November 29-December 4
	60 degrees	December 18-23
	90 degrees	June 11-18
	90 degrees	November 6-13
	120 degrees	June 26-July 2
	120 degrees	October 18-26
	180 degrees	July 27-August 6
Mercury-Jupiter	60 degrees	February 15-19
	60 degrees	June 29-July 13
	0 degrees	March 23-April 1
	90 degrees	January 14-22
	90 degrees	July 13-20
	120 degrees	August 2-19
	120 degrees	August 23-September 11
	120 degrees	September 17-29
	120 degrees	December 1-23
	180 degrees	October 24-November 1

Planetary aspect cycles pre-date the work of Jeanne Long. In 1946, astrologer Donald Bradley $_{(2)}$ defined a model based on geocentric pairings of planets. In his model, as a pair of planets approach one another and come to within 15-degrees separation, a sinusoidal weighting is applied to the separation. At a 15-degree separation the weighting assigned is zero. At a 0-degree separation (planets are conjunct) the weighting is 10. Bradley's model also included a variable defined as the mathematical average of the declination of Venus and Mars. Running the model on a daily basis generates a plot with many inflection points. The image in Figure 10-5 illustrates the Bradley Model plot for 2023.

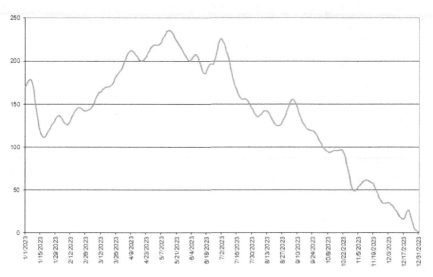

Figure 10-5
Bradley Model Plot

For the past several years, the Bradley Model has been largely ineffective owing to the injection by the Federal Reserve of over $100 billion dollars per month into the financial system. However, in early 2022 when the Federal Reserve announced it would be curtailing its liquidity injections, the model started aligning once again with the price behavior on the S&P 500.

Many people mis-interpret the Bradley Model. They take the slope of the model to mean the S&P 500 price trend will have the same slope. That is, they feel the market should rise when the Bradley plot is sloping higher. I have come to realize that the most effective way to interpret the model is to focus on the inflection points in the output plot. An inflection point, when the slope of the plot suddenly changes, has a very strong correlation to price changes on the S&P 500. I take this as evidence that planetary pairings affect the human emotions of fear and greed, both of which propel financial markets. For example, the model expressed an inflection point in late December 2021. This point aligned to the peak of the S&P 500. The model expressed an inflection point on June 16. The S&P 500 recorded a low on June 15. The S&P 500 recorded a swing high point on August 16, 2022. The Bradley model displays an inflection point on August 18. The model displays inflection points at November 5, November 16, and a broader inflection between December 4-8, 2022.

For 2023, the following dates are the inflection points on the model plot:

January 6, 16
February 2, 9, 20, 25
March 17
April 9, 19, 29
May 5, 14
June 2, 8, 16,22
July 2
August 4, 12, 24
September 6
October 9, 21
November 1, 12
December 17, 22.

CHAPTER ELEVEN
Shemitah and Kabbalah Cycles

Just as intriguing as synodic and sidereal cycles, are cycles rooted in religious doctrine that intersect with financial market turning points. One religious concept is that of Shemitah which is rooted in the Hebrew Bible. I first learned of the Shemitah from the writings of Rabbi Jonathan Cahn. [1][2][3] On the surface, Cahn appears to be an average, ordinary Rabbi from New Jersey, USA. But behind the scenes, he has done a masterful job of applying Shemitah to the financial markets. His books include: *The Harbinger, The Book of Mysteries,* and *The Paradigm.*

As Cahn explains, in the book of Exodus (Chapter 23, verses 10-11), it is written: *You may plant your land for six years and gather its crops. But during the seventh year, you must leave it alone and withdraw from it.*

In the book of Leviticus (Chapter 25, verses 20-22), it is written: *And if ye shall say: "What shall we eat the seventh year? Behold, we may not sow, nor gather in our increase"; then I will command My blessing upon you in the sixth year, and it shall bring forth produce for the three years. And ye shall sow the eighth year, and eat of the produce, the old store; until the ninth year, until the produce come in, ye shall eat the old store.*

Breaking these Biblical statements down into simple-to-understand terms means that every 7th year something will happen in the geopolitical economy and on the financial markets.

The first Shemitah year in the modern State of Israel was 1951-52 Subsequent Shemitah years have been 1958–59, 1965–66, 1972–73, 1979–80, 1986–87, 1993–94, 2000–01, 2007–08, 2014-15, and 2021-2022. The next Shemitah year will be 2028-2029.

Shemitah years can be grinding and difficult for market participants. The most recent Shemitah year concluded in September 2022. During this Shemitah year, Russia invaded Ukraine; the price of Oil surged; the price of gasoline at the pump surged; fertilizer prices for farm operators surged, food prices surged; inflation reached the 8% level; the Federal Reserve ceased its liquidity injections into the financial system; BitCoin got pumelled, and the equity markets sagged from January through September.

A Shemitah year starts in the month of Tishrei (the first month of the Jewish civil calendar) and ends in the month of Elul. This timeframe on the Gregorian calendar will roughly run from September to the following September. The website www.chabad.org will help identify the exact dates.

The chart in Figure 11-1 illustrates the E-mini S&P 500 with some recent Shemitah years overlaid. As Figure 11-1 further suggests, the time immediately following the conclusion of a Shemitah year can bring further drama. Equity markets can move modestly lower in the aftermath of a Shemitah year. The earlier-cited Biblical passages advise "... *ye shall sow the eighth year, and eat of the produce, the old store; until the ninth year, until the produce come in, ye shall eat the old store.*" In other words, do not expect a bounty in the year immediately after a Shemitah year. The post-Shemitah period for the 2021-2022 Shemitah year will conclude by September 2023.

Shemitah years are not always solely about the equity markets. The chart in Figure 11-2 illustrates Oil prices with Shemitah years overlaid.

Figure 11-1
E-mini S&P 500 and Shemitah years

Those who understood and used Shemitah as an investing timing strategy made serious money on the Crude Oil market in 2007, 2014, and 2022.

Figure 11-2
Crude Oil and Shemitah years

According to Rabbi Cahn, in addition to the Shemitah year, there are certain dates from the Hebrew calendar that can have a strong propensity to align with swing highs and lows on the New York Stock Exchange. Rabbi Cahn advises to pay close attention to four particular dates from the Hebrew calendar:

- ✿ The 1st Day of the month of Tishrei marks the start of the Jewish civil calendar, much like January 1 marks the start of the Gregorian calendar.

- ✿ The 1st day of the month of Nissan marks the start of the Jewish sacred year.

- ✿ The 3rd important date is the 9th day of the month of Av which marks the date when Babylon destroyed the Temple at Jerusalem in 586 BC. Other calamitous events have beset the Jewish people on the 9th of Av throughout history. In particular, Cahn tells of the mass expulsion of Jewish people from Spain in 1492. As this expulsion was going on, a certain explorer with three ships was about to set sail on a voyage of discovery. That explorer sailed out of port on August 3, 1492 which was one day after the 9th of Av. That explorer was Christopher Columbus.

- ✿ The 4th key date in the Jewish calendar is Shemini Atzeret (the Gathering of the Eighth Day). This date typically falls somewhere in late September through late October in the month of Tishrei.

The website www.chabad.org allows one to quickly scan back over a number of years to pick off these important dates. My back-testing has shown a variable correlation to these dates. But the correlation is valid enough that I recommend traders and investors pay attention to these dates.

For 2022, these critical dates fell as follows: 1st of Nissan on March 26, the 9th of Av on July 30, the 1st of Tishrei on September 19, and Shemini Atzeret on October 10.

March 26 landed on a Saturday. On the following Tuesday (March 29), the S&P 500 stopped a rally and resumed its downtrend. The 9th of Av date did not affect the market. From September 13 through 16, the S&P 500 declined. The 19th of September provided a move higher and a glimmer of hope that the downtrend might end. However, it was not to be. The very next day the downtrend resumed in earnest. October 10 fell on a Saturday. On Monday, October 12, the S&P staged an intra-day reversal of trend and embarked on a counter-trend rally.

For 2023, these critical dates will fall as follows:

- ✿ **1st of Nissan on March 23**
- ✿ **9th of Av on July 27**
- ✿ **1st of Tishrei on September 16**
- ✿ **Shemini Atzeret on October 7.**

In addition to cycles related to religious doctrine, cycles related to Kabbalah mathematics are also intriguing. A 2005 article in *Trader's World* magazine (4) suggested that W.D. Gann might have been instilled with knowledge of Jewish mysticism based on Kabbalah doctrine. Gann apparently had connections to a New York personality named Sepharial who is said to have taught Gann about astrology and esoteric matters.

The Kabbalah centers around the Hebrew Alef-bet (alphabet). The Hebrew Alef-bet comprises 22 letters. In Kabbalistic methodology, these letters are assigned a numerical value. Starting with the first letter, values are 1, 2, 3, 4, 5, 6, 7, 8, 9, 10, 20, 30, 40, 50, 60, 70, 80, 90, 100, 200, 300, and 400.

There are many mathematical techniques that can be applied to parsing the Alef-bet. One in particular involves taking the odd-numbered letters and the even numbered letters and assigning their appropriate numerical values.

The numerical value (sum total) of the Alef-bet is 1495. The sum total of the odd-numbered letters is 625. The sum total of the even numbered letters is 870.

- ✿ 625/1495 = 42%. Taking a circle of 360-degrees, 42% is 150.5 degrees.

- ✿ 870/1495 = 58%. Taking a circle of 360-degrees, 58% is 209.5 degrees.

- ✿ Kabbalists are also well aware of phi as it pertains to the Golden Mean. Phi is famously known as 1.618.

- ✿ 1/phi = 62%. Taking a circle of 360-degrees, 62% is 222.5 degrees.

- ✿ 1 − (1/phi) = 48%. Taking a circle of 360-degrees, 48% is 137.5 degrees.

From a significant price low (or high) starting point, one can examine price charts for time intervals when a geocentric or heliocentric planet advances 137.5, 150.5, 209.5, or 222.5-degree amounts. My back-testing has shown that Venus and Mars heliocentric advances are very apt to align to market turning points.

To illustrate, consider the significant low in March 2009 on the S&P 500 as a start point. Applying the above described 137.5-degree ratio, what emerges is the discovery that 11 years after the March 2009 lows the 137.5-degree advancements of heliocentric Venus land spot on the March 2020 lows.

Extending these various degree intervals forward from the 2009 lows reveals a series of correlations to swing points on the S&P 500 chart some thirteen years later.

For example, Figure 11-3 shows how the 209.5-degree intervals landed at a January 2022 swing low, a June 2022 swing high, and an October 2022 severe intra-day reversal.

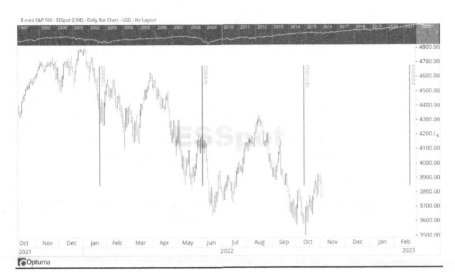

Figure 11-3
E-mini S&P 500 and Venus 209.5-degree intervals

To further illustrate the effect of these intervals, consider that in November 2008 the Nasdaq recorded a low that was slightly less than the March 2009 low.

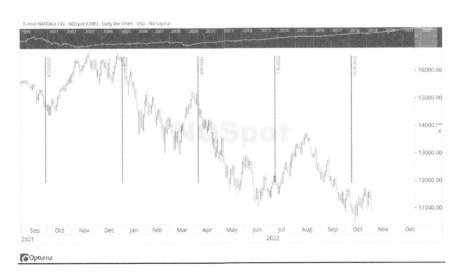

Figure 11-4
Nasdaq and Venus 150.5-degree intervals

Using the November 2008 date as a start point and applying 150.5-degree Venus intervals shows correlations still exist in 2022 as shown in Figure 11-4.

Application of these intervals is not limited to equity markets, nor is Venus the only planet to consider. Consider the 2011 significant high for Gold. Figure 11-5 illustrates the application of 137.5, 150.5, and 209.5-degree intervals of heliocentric Mars from the 2011 high.

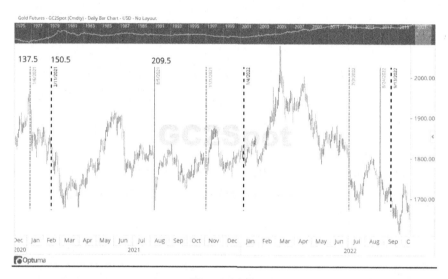

Figure 11-5
Gold and Mars intervals

The gold price reversal in January 2021 aligns to a 137.5-degree interval. A swing low in August 2021 aligns to a 209.5-degree interval. A steep drop in July aligns to a 137.5-degree interval.

Consider also the 2008 price highs on Oil. Extending Mars 137.5-degree intervals from those highs reveals that an interval landed right at the start of the steep price decline that briefly took Oil prices negative in 2020. Another interval lands mere days after the March 2022 price highs at $130 per barrel.

Consider too, the Soybean market. From a significant price low in 2019, heliocentric Mars intervals of 137.5, 150.5, and 209.5-degrees can be seen aligning to various swing highs and lows.

To assist you in identifying these heliocentric intervals:

- ☼ **For the S&P 500 in 2023 (using a start point of March 2009), the 209.5-degree intervals of Venus to watch for will occur: February 21, June 12, and October 22.**

- ☼ **For the Nasdaq in 2023 (using a start point of the November 2008 lows), the 150.5-degree intervals of Venus to watch for will occur: January 12, April 15, July 17, and October 20.**

- ☼ **For Oil in 2023 (using a start point of July 14, 2008) the 137.5-degree interval of Mars to watch for will occur at: August 16.**

- ☼ **For Gold in 2023 (using a start point of September 6, 2011), the Mars degree intervals to watch for will occur: March 10 (137.5), July 22 (150.5), and October 27 (209.5), and December 30 (137.5).**

- ☼ **For Soybeans in 2023 (using a start point of May 10, 2019), the Mars degree intervals to watch for will occur: April 27 (150.5), October 2 (137.5), and October 28 (209.5).**

- ☼ **For the US Dollar Index in 2023 (using a start point of 21 April 2008), the Venus degree intervals to watch for will occur: March 20, June 21, September 24, December 27 (150.5), February 17, July 5, November 24 (222.5), February 8, May 4, July 29, and October 23 (137.5).**

CHAPTER TWELVE
Lunar and Planetary Declination

The Moon's orbit around the Earth can be described in terms of either synodic or sidereal cycles. The sidereal period of the Moon is 27.5 days (as viewed from a fixed reference like the Sun) and the synodic period is 29.53 days (as viewed from our vantage point here on Earth). This latter period lends itself to the expression *lunar month*.

During each synodic lunar cycle, the Moon can be seen to vary in its position above and below the lunar ecliptic; it will go from maximum declination to maximum declination.

Moon declination takes on an intriguing aspect when one considers that in 2020 two of the Moon declination minima events occurred on February 19 and March 18. The exact trend change turning point ahead of the COVID panic sell-off came at February 19. The panic lows came at March 23, mere days after the declination low.

It is said that W.D. Gann was a proponent of following lunar declination when trading Soybeans and Cotton. It is likely that he appreciated how the Moon's gravitational pull governs the ocean tides. With our bodies being substantially water, he was likely postulating that the Moon was influencing the emotions of Soybean and Cotton traders.

Figure 12-1 presents a chart of Soybeans with the Moon declination in the lower panel. The vertical lines overlaid on the chart illustrate the propensity for declination maxima, minima, and zero points to align to price swings. A 4-hour chart would be the ideal way to study these price swings closely. In Mr. Gann's day, he had access to ticker tape data which allowed him to accurately study short term price moves as Moon was nearing its declination minima, maxima, and zero declination point.

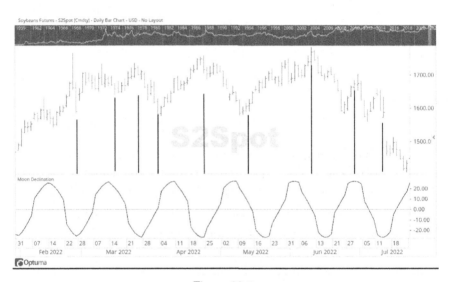

Figure 12-1
Moon Declination and Soybeans

Figure 12-2 illustrates Cotton futures prices during 2022 to date. I have overlaid the chart with a dark vertical line at some maximum, minimum, and zero-point lunar declination events. You can now appreciate why a trader like W.D. Gann would have used lunar declination phenomenon to give himself an advantage in the Cotton and Soybean markets.

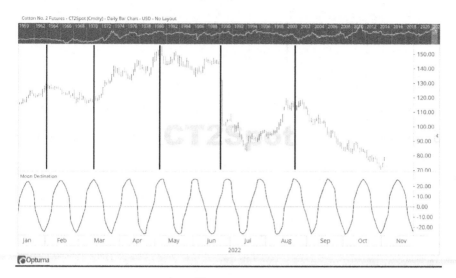

Figure 12-2
Moon Declination and Cotton

Figure 12-3 illustrates Moon declination and the S&P 500. Maximum, minimum, and zero levels of declination all bear a strong correlation to swing points in price.

Figure 12-4 illustrates Moon declination and Gold futures. Maximum and minimum levels of lunar declination bear a strong correlation to swing pivots in price.

Although not shown here, it is worth noting also that in March 2022, the price of Oil peaked at near $130 per barrel within two days of a Moon maximum declination event. As well, in February 2022, Wheat prices began a sizeable rally within two days of a Moon minimum declination event.

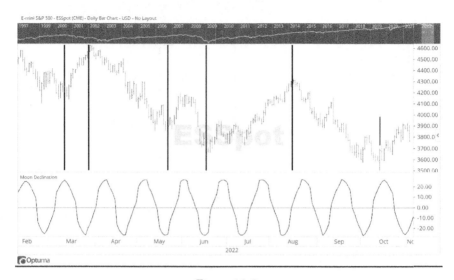

Figure 12-3
Moon Declination and E-mini S&P 500

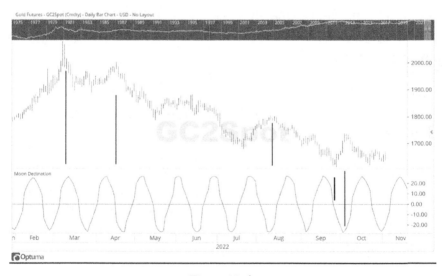

Figure 12-4
Moon Declination and Gold

Moon declination events also align very closely to some individual stocks. Consider the favorite stocks you enjoy trading in and out of. Does Moon declination align to periodic price swing points?

To assist you with some back-testing of your own, consider that in 2021:

Moon was at its maximum declination: January 1, January 27, February 23, March 22, April 19, May 16, June 12, July 9, August 6, September 2, September 29, October 27, November 23, December 20.

Moon was at its minimum declination: January 13, February 9, March 8, April 4, May 1, May 29, June 26, July 23, August 20, September 15, October 12, November 9, and December 6.

Moon was at zero declination: January 6, January 19, February 2, February 15, March 1, March 14, March 28, April 11, April 25, May 8, May 22, June 4, June 18, July 1, July 15, July 28, August 12, August 25, September 9, September 21, October 6, October 19, November 2, November 14, November 29, December 13, and December 27.

To further assist you, consider that **in 2022**:

Moon was at its maximum declination: January 17, February 13, March 12, April 8, May 6, June 2, July 26, August 23, September 19, October 15, November 13, and December 10.

Moon was at its minimum declination: January 30, February 26, March 25, April 22, May 19, June 15, July 13, August 9, September 6, October 3, October 30, November 26, and December 24.

Moon was at zero declination: January 8, January 23, February 5, February 20, March 5, March 19, April 1, April 15, April 28, May 12, May 25, June 9, June 21, July 5, July 19, August 3, August 15, August 29, September 11, September 26, October 9, October 23, November 5, November 19, December 2, December 17, and December 30.

For 2023:

Moon will be at its maximum declination: January 6, February 2, March 2, March 28, April 25, May 22, June 18, July 15,

August 12, September 8, October 5, November 2, November 29, and December 26.

Moon will be at its minimum declination: January 20, February 16, March 15, April 12, May 9, June 5, July 2, July 30, August 27, September 24, October 20, November 16, and December 14.

Moon will be at zero declination: January 13, January 26, February 9, February 23, March 8, March 22, April 5, April 18, May 2, May 15, May 29, June 11, June 26, July 8, July 23, August 5, August 19, September 1, September 15, September 29, October 13, October 26, November 9, November 23, December 6, and December 20.

Individual planets also experience declinations of up to about 25 degrees above and below the celestial equator plane. My introduction to planetary declination came several years ago with the discovery of a 25-year-old astrology book in a used bookshop.[1] The book used the method of parallel and contra-parallel declinations using the declination levels that were in place at the date the stock or commodity first started trading on an exchange (first trade date). This approach proved tedious to work through, even though the results were favorable.

My research has now pointed me to a more elegant approach apparently used by W.D. Gann in his day. In his 1927 book, *Tunnel Through the Air*, [2] Gann references days when the story hero Robert Gordon was very certain trend changes would occur on his Major Motors stock trades. Gann hints strongly that instead of watching for two planets to become parallel or contra-parallel so as to match the situation at the first trade date, one should instead be watching for Mars and Venus to pass the same declination level as they were at on the first trade date (their natal declination levels). When I checked the Mars and Venus planetary declination at the Major Motors first trade date and compared those figures to Robert Gordon's key dates, I found a correlation.

Dates when Venus and Mars pass through their natal declination levels are critical dates to pay attention to. Consider the following examples of Gold, Cotton, Crude Oil, Soybeans, and the NYSE:

Gold

On December 31, 1974 when the Gold futures instrument was launched, Venus was -22.42 degrees declination and Mars was -22.45 degrees.

In 2021, Mars and Venus passed through their December 31, 1974 natal declination levels (+/- 1.5 degrees tolerance) as follows:

- ✪ January 1 through January 30 (Venus)
- ✪ October 1 through 10 (Venus)
- ✪ December 22, 2021 to March 7, 2022 (Mars).

Gold prices peaked on January 6 and proceeded to decline by nearly $300 per ounce. September 29 was a swing pivot low that saw Gold move higher by $150 per ounce in the weeks that followed. Gold moved higher in price at year end and into early 2022. Curiously enough, Gold prices peaked at over $2000 per ounce at March 7.

In 2022, Venus passed through its December 31, 1974 natal declination level from November 20 to December 31.

For 2023 Mars and Venus will be at their December 31, 1974 natal declination levels:

- ✪ **January 1-6 (Venus)**
- ✪ **November 26-December 23 (Mars).**

Cotton

W.D. Gann was reported to have frequently traded Cotton futures. Cotton futures started trading in New York on June 20, 1870. At that date, Venus was +14.84 degrees of declination and Mars was +21.42 degrees of declination.

In 2021, Mars was at its 1870 first trade natal declination from February 20 to March 15 and again from May 28 to June 21. Venus was at its 1870 natal declination from April 20-28 and from July 15-22.

In 2021, Cotton futures exhibited an interim high on February 24; a price rally started on May 27. April 27 marked an interim high.

In 2022, an 11 cent per pound drawdown appeared as Venus was passing its 1870 declination.

For 2023, Mars and Venus will be at their June 1870 natal declinations:

- ☿ **March 21 to 28, and June 25 to July 5 (Venus)**
- ☿ **May 16-31 (Mars).**

Crude Oil

WTI Crude Oil futures started trading in New York on March 30, 1983. At that date, Venus was +16.28 degrees of declination and Mars was +9.52 degrees of declination.

In 2021, Venus was its natal declination level (+16.28 degrees) from April 24 to May 2 and July 8 to 19. Mars was its natal declination level (+9.52 degrees) August 5 to 17. Oil exhibited a $15 per barrel drawdown starting right at the July 8 date. Oil turned in a significant swing low a mere three days after the August 17 date.

In 2022, Venus was at its natal declination level (+16.28 degrees) from June 12 to 21 and August 22-30. Mars was at its natal declination level (+9.52 degrees) from June 29-July 10. On June 14, Oil prices peaked at just over $120 per barrel. June 29 saw the start of another down leg which took Oil prices back beneath $100 per barrel. In late August, Crude Oil attempted a counter-trend rally but the rally failed right at August 30. Oil prices then dropped nearly $20 per barrel.

For 2023:

- ☼ **Venus will be at its 1983 first trade declination level March 26-29 and again June 23-29**

- ☼ **Mars will be at its 1983 first trade declination level July 16-28.**

Soybeans

The Chicago Board of Trade was founded April 3, 1848. W.D. Gann was very cognizant of this date when trading Soybeans. Looking at Soybean futures through the lens of this date, as opposed to the 1936 date when Soybean futures actually started trading, yields some interesting finds.

At the 1848 date, Venus was at -7.26 degrees of declination and Mars was at +24.8 degrees of declination.

For 2021, Venus was at its natal declination level (-7.26 degrees) from March 6-12, and again from August 29-September 4. The March timeframe marked a $1 per bushel price drawdown.

Mars was at its natal declination level (+24.80 degrees) from March 17-May 24. A significant price high reversal in early May 2021 aligned to the Mars declination.

In 2022, Venus passed its natal declination level (-7.26 degrees) between April 12-21. A drawdown of over $1 commenced right at the April 21 date. Venus again passed its natal declination level between October 14-20. During this time, Soybeans recorded a swing bottom and began moving up in price.

In 2022, Mars passed its natal declination level (+24.80 degrees) between October 20 and year end.

In October 1936 when the Soybeans contract first started trading in Chicago, Mars was at +10.32 degrees of declination and Venus was at -15.06 degrees of declination. In 2022, Mars passed through this declination level between July 2 and 12. On July 12, Soybean prices suddenly dropped; in fact, gapping lower on the chart by over $1 per bushel.

Venus passed through -15.06 degrees of declination between January 20-February 5, March 13-23, and November 1-7. During the January and March periods, Soybeans gained $1 per bushel each time. At this time of writing, the November date seems to be aligning to a push higher in prices.

For 2023:

- ☼ **Venus will be at the -7.26 degree level of natal declination from February 5-10, and September 21-October 4**

- ☼ **Mars will be at the +24.80 degree level of natal declination from January 1 through May 1.**

NYSE

The New York Stock Exchange (NYSE) can also be viewed through the lens of declination. As the next chapter will reveal, the NYSE traces its origins to May 17, 1792. At that date Venus was at +11.42 degrees declination and Mars was at +5.30 degrees of declination.

It is intriguing to learn that in late February, 2020 full-blown COVID panic was setting in across equity markets. Venus was approaching its natal declination level of +11.42 degrees declination.

In 2021, Mars was at its 1792 first trade declination from August 24-30. In the immediate aftermath of this declination event, the S&P 500 declined 6%.

In 2022, Venus passed its 1792 natal declination level from May 30-June 5 and again from September 3-11. May 30 marked the failure of a small counter-trend rally. On September 6, another counter-trend rally started which pushed the S&P 500 up by nearly 200 points.

Mars passed its natal declination level between June 16 and 21. At this time of writing, it can be reported that June 16 marked a swing low that resulted in the S&P 500 rallying near 300 points.

For 2023:

    ✪ **Mars will pass its natal declination level from August 3-14th**

    ✪ **Venus will pass its natal declination level on three occasions, March 13-20, July 9-15, and September 2-October 4.**

W.D. Gann was very generous to leave us the Mars and Venus clue in Tunnel *Through the Air* . Historical declination data is readily available on the internet. Or, a software program such as *Solar Fire Gold* will also help you identify the data.

In chapter 14, a variety of commodity futures products will be discussed. Declination levels will be provided as part of each individual discussion.

CHAPTER THIRTEEN
NYSE 2023 Astrology

In our hyper-linked economy, events on the New York Stock Exchange can quickly reverberate across other global exchanges. It is for this reason that all of my Almanacs to date have focused on the New York Exchange.

No examination of the astrology of the New York Stock Exchange would be complete without mention of Louise McWhirter. After years of reading old papers and manuscripts, I still have no idea who Louise McWhirter was. What I do know is, Louise McWhirter focused not only on the 18.6 year cycle, but also on the astrology of the New York Exchange. Her technique, which revolves around the New Moon (lunation), remains viable to this day.

One observation of interest pertains to the actions of the Federal Reserve in the aftermath of the COVID 19 panic selloff. Massive amounts of fiat liquidity were pumped into the banking system to lift the economy and the markets. This liquidity partly obscured any weakness on equity markets with 'buy the dip' being the resounding cry when markets expressed weakness. The Federal Reserve reigned in its liquidity injections in early 2022 to quell inflation. Almost immediately, astrology in general and the McWhirter method in particular started to correlate to market movements again.

The Lunation and the New York Stock Exchange

Lunation is the astrological term for a New Moon. At a lunation, the Sun and Moon are separated by 0-degrees and are together in the same sign of the zodiac. The correlation between the monthly lunation event and New York Stock Exchange price movements was first popularized in 1937 by McWhirter. In her book, *Theory of Stock Market Forecasting,* [1] she discussed how a lunation exhibiting hard aspects to planets such as Mars, Jupiter, Saturn, and Uranus was indicative of a lunar cycle during which the New York Stock Exchange would display notable volatility. She further said that hard aspects to the natal outer planets in the NYSE 1792 horoscope should be watched.

She also paid close attention to Mars and Neptune, the two planets that rule the New York Stock Exchange. The concept of planetary rulership extends back to the 1800s when astrologers began denoting each zodiac sign as having a celestial body that ruled that sign. The planetary placements at the day and time the New York Stock Exchange was founded in 1792 were such that Mars and Neptune were the rulers of the zodiac at that moment. This is because the 10th House of the NYSE birth horoscope spans the signs of Pisces and Aries. Neptune rules Pisces and Mars rules Aries. McWhirter said those times of a lunar month when the transiting Moon makes 0-degree aspects to Mars and Neptune should be watched carefully as volatility could result.

New York Stock Exchange – First Trade Chart

The New York Stock Exchange officially opened for business on May 17, 1792. As the horoscope in Figure 13-1 shows, the NYSE has its Ascendant (Asc) at 14-degrees of Cancer and its Mid-Heaven (MC) at 24-degrees of Pisces.

McWhirter further paid close attention to those times in the monthly lunar cycle when the transiting Moon passed by the NYSE natal Asc and MC locations at 14 Cancer and 24 Pisces respectively.

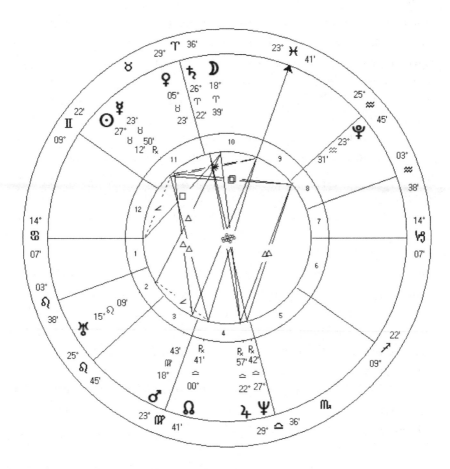

Figure 13-1
NYSE First Trade horoscope

The McWhirter Methodology

In my research and writing, I follow the McWhirter methodology for shorter-term trend changes. When forecasting whether or not a coming month will be volatile or not for the NYSE, the McWhirter methodology starts with creating a horoscope chart for the New Moon date and positioning the Ascendant of the chart at 14-degrees Cancer (the Ascendant position on the 1792 natal chart of the New York Stock Exchange). Positioning the Ascendant is made easy with clicks of the mouse when using the *Solar Fire Gold* software program. Aspects to the lunation are then studied. If the lunation is at a 0, 90, or 120-degree aspect to Mars, Neptune, 14 Cancer, or 24 Pisces, one can expect a volatile month ahead. A lack of such aspects portends a less volatile period. The McWhirter method further demands a consideration of where the Moon is at each day. Aspects of the transiting Moon to Mars, Neptune, 14 Cancer, or 24 Pisces represent dates of potential short-term trend reversals. Although not expressly stated by McWhirter, it is also important to pay attention to those dates when Moon is at either maximum or minimum declination. As well, dates when Mercury is retrograde and dates when Venus is at or near its maximum or minimum declination should be considered carefully.

Similarly, when studying an individual stock or an individual commodity futures contract, the McWhirter approach calls for the creation of a horoscope chart at the first trade date of the stock or commodity. The Ascendant is then shifted so that the Sun is at the Ascendant. The software program *Solar Fire Gold* is very good for generating first trade horoscope charts for McWhirter analysis where the Ascendant needs to be shifted. As to why she placed the Sun at the Ascendant for an individual stock and 14 of Cancer at the Ascendant for analyzing a lunation event, remains unclear to me.

In stock and commodity analyses, McWhirter paid strict attention to those times of a calendar year when transiting Sun, Mars, Jupiter, Saturn, Neptune, and Uranus made hard 0, 90, and 180-degree aspects to the natal Mid-Heaven, natal Ascendant, natal Sun, natal Jupiter and even the natal Moon of the individual stock or commodity future being studied.

One must be alert at these aspects for the possibility of a trend change, the possibility of increased volatility within a trend, or even the possibility of a breakout from a chart consolidation pattern. Evidence of such trend changes will be found by applying chart technical indicators as discussed in an earlier chapter.

February-March 2020: an Historical Example

The McWhirter method can be thoroughly appreciated by examining events around the 2020 COVID panic sell-off:

- ✿ A New Moon on February 23, 2020 came during Mercury retrograde

- ✿ Within a day of the New Moon, the Moon transited past both the NYSE natal Mid Heaven (24 Pisces) and the co-ruler Neptune

- ✿ A day later, Moon transited past co-ruler Mars. Saturn was 90-degrees square the NYSE 1792 natal Saturn point

- ✿ Jupiter was 120 degrees (trine) to the 1792 natal Mars point. Sun was conjunct the 1972 natal Pluto point

- ✿ This was a lot of planetary energy compressed into a short timeframe. The NYSE reacted in a strongly negative manner.

- ✿ An attempt to counter the negative sentiment failed as Moon made a hard 90-degree aspect to Neptune and a 0-degree aspect to 14 of Cancer on March 2 and 5 respectively

- ✿ On March 24, a New Moon event again arrived which was within 5-degrees of being conjunct to NYSE co-ruler Neptune. With that bit of positive planetary energy, the markets began a sharp recovery.

As a trader or investor, you can view the events of February-March, 2020 through the lens of the Federal Reserve and fiat stimulus or you can view the events through the lens of the New Moons and aspects to key NYSE natal zodiac points.

What follows in this chapter is a listing of the date for each lunar cycle in 2023 along with a list of times when Moon passes Mars, Neptune, the

NYSE natal Mid-Heaven at 24 Pisces, and the NYSE natal Ascendant at 14 Cancer. In addition, dates of planetary retrograde, dates of declination maxima and minima, and dates of planetary elongation are included. Also included is a lunar feature called *Moon Void of Course*. Moon is considered Void of Course (VOC) from the time it makes no aspects to other planets to the time it enters the next zodiac sign. In a given month, the Moon will be VOC approximately 12 times. In order for VOC to affect equity markets, the VOC event must occur Monday through Friday, must be more than four hours in duration, and must occur during NYSE trading hours. I disregard any VOC events outside these parameters. The net result is that in a typical month there might be up to four VOC events.

Lastly, I have included dates in each lunar cycle when Sun, Mars, Saturn, and Jupiter make aspects to the NYSE 1792 natal Moon, natal Sun, and the various natal outer planets.

2023 Lunation Events

December 2022 - January 2023

Market action in January 2023 will be influenced by the New Moon cycle that commenced on December 23, 2022 (Sun at 2 Capricorn) and runs to January 21, 2023 when Sun will be at 1 Aquarius. At this lunation, Sun, Moon, Venus, Mercury, and Pluto are all compressed into one sign. Collections of planets in close proximity to one another can often have consequences. This lunar cycle is punctuated by Mercury being retrograde. The New Moon is at a 90-degree aspect to expansive Jupiter which could indicate a trend change on the market. Mars at its declination maximum is also poised to play a trend changing role. Lastly, an aspect from December 2022 that continues into 2023 could play a trend-changing role. The aspect is a 45-degree semi-square between Jupiter and Uranus.

Key dates during this lunar cycle are:

- ☼ December 29: Moon passes Neptune and the NYSE natal mid-Heaven point at 24 Pisces

- ☼ December 29: Mercury turns retrograde

- ☼ January 3: Mercury at Perihelion

- ☼ January 3: Moon passes NYSE co-ruler Mars

- ☼ January 6: Moon at declination maximum

- ☼ January 6: Moon passes 14 of Cancer

- ☼ January 6: Saturn 120-degrees to NYSE natal Jupiter

- ☼ January 9: Sun 120-degrees trine NYSE natal Mars

- ☼ January 13: Sun 90-degrees square NYSE natal Jupiter

- ☼ January 16: Sun 90-degrees square NYSE natal Saturn

- ☼ January 17: Moon VOC

- ☼ January 17: Sun 120-degrees trine NYSE natal Sun

- ☼ January 17: Sun 90-degrees trine NYSE natal Neptune

- ☼ January 19: Moon VOC

- ☼ January 25-26: FOMC (Federal Open Market Committee) meeting

- ☼ January 26: Moon at its minimum declination.

January- February 2023

Market action through late January and much of February, 2023 will be influenced by the New Moon cycle that commences on January 21, 2023 (Sun at 1 Aquarius) and runs until February 20 (Sun at 1 Pisces). The lunation is a favorable 60-degrees to Jupiter and 120-degrees to Mars. This lunar cycle contains two particular dates of interest: Mercury marking its greatest westerly elongation and Mercury ending its retrograde movement. Either of these could initiate a trend change on the NYSE.

Key dates during this lunar cycle are:

- ☼ January 23: Moon VOC
- ☼ January 24: Moon passes Neptune and the NYSE natal mid-Heaven point at 24 Pisces
- ☼ January 30: Mercury at its greatest westerly elongation
- ☼ January 30: Moon passes NYSE co-ruler Mars
- ☼ February 1: Moon VOC
- ☼ February 2: Moon passes 14 of Cancer
- ☼ February 2: Moon at maximum declination
- ☼ February 6: Moon VOC
- ☼ February 11: Sun 120-degrees trine NYSE natal Jupiter
- ☼ February 12: Saturn 90-degrees square NYSE natal Sun
- ☼ February 15: Sun 90-degrees trine NYSE natal Sun
- ☼ February 16: Moon at declination minimum
- ☼ February 16: Saturn 120-degrees trine NYSE natal Neptune
- ☼ February 16: Sun 120-degrees trine NYSE natal Neptune
- ☼ February 16: Mercury at Aphelion
- ☼ February 17: Mercury retrograde ends

February - March 2023

Market action through late February and much of March 2023 will be influenced by the New Moon cycle that commences on February 20, 2023 (Sun at 1 Pisces) and runs until March 21 (Sun at 0 Aries). This lunation is conjunct heavy-weight planet Saturn. This conjunction is suggestive of some added volatility. Possibly adding to the volatility will be a 30-degree aspect between Jupiter and Uranus. Key dates during this lunar cycle are:

Key dates during this lunar cycle are:

- ✿ February 21: Moon passes co-ruler Neptune and natal mid-Heaven point of 24 Pisces

- ✿ February 27: Mars 90-degrees square NYSE natal Mars.

- ✿ February 28: Moon passes co-ruler Mars

- ✿ March 1: Moon at declination maximum

- ✿ March 2: Moon passes 14 Cancer

- ✿ March 4: Sun 120-degrees trine NYSE natal Ascendant

- ✿ March 10: Mars 90-degrees square NYSE natal Mid-Heaven

- ✿ March 14: Sun conjunct NYSE natal Mid-Heaven

- ✿ March 13-20: Venus declination matches its 1792 natal NYSE declination

- ✿ March 15-16: FOMC meets

- ✿ March 15: Moon at declination minimum

- ✿ March 15: Jupiter 120-degrees trine to NYSE natal Uranus

- ✿ March 20: Mars 120-degrees trine to NYSE natal Neptune.

March - April 2023

The April New Moon cycle commences on March 21, 2023 (Sun at 0 Aries). This lunar cycle runs until April 20, 2023 (Sun at 29 Aries). The lunation itself is within orb of being at a 0-degree hard aspect to the NYSE natal Mid-Heaven and co-ruler Neptune. The lunation is also within orb of being at a hard 90-degree aspect to NYSE co-ruler Mars. These aspects could energize this cycle. In addition, the first two weeks of April will have Sun and Jupiter making a conjunction. This combination has a propensity for trend changes.

Key dates during this lunar cycle are:

- ✿ March 28: Moon at declination maximum
- ✿ March 28: Moon passes NYSE co-ruler Mars
- ✿ March 28: Jupiter conjunct NYSE natal Moon
- ✿ March 29: Moon passes 14 of Cancer
- ✿ March 29: Moon at maximum declination
- ✿ April 3: Sun 90-degrees square NYSE natal Ascendant
- ✿ April 4: Moon VOC
- ✿ April 4: Mercury at Perihelion
- ✿ April 6: Moon VOC
- ✿ April 8: Sun 120-degrees trine natal Uranus
- ✿ April 11: Mercury at greatest easterly elongation
- ✿ April 11: Moon VOC
- ✿ April 12: Moon at declination minimum.
- ✿ April 16: Sun conjunct NYSE natal Saturn
- ✿ April 17: Moon passes NYSE natal Mid-Heaven point.

April - May 2023

The May New Moon cycle commences on April 20, 2023 (Sun at 0 Taurus). This lunar cycle runs until May 19, 2023 (Sun at 27 Taurus). The lunation is just within orb of being at a favorable 60-degree aspect to NYSE co-ruler Mars. However, Mars sits atop the NYSE natal Ascendant at 14 Cancer. This sets up competing cosmic energy which could be reflected through market volatility. Mercury turns retrograde one day into this lunar cycle which could further add to market volatility. As well, Jupiter, Sun, Node, Moon, Uranus, and Mercury are all squeezed into a 30-degree portion of the zodiac. Compressed planetary energies can have consequences.

Key dates during this lunar cycle are:

- ✿ April 21: Mercury turns retrograde
- ✿ April 21: Mars conjunct the NYSE natal Asc at 14 Cancer
- ✿ April 23: Venus approaches its declination maximum and Mars completes its declination maximum
- ✿ April 24: Moon VOC
- ✿ April 24: Moon at declination maximum
- ✿ April 25: Moon passes 14 of Cancer and also NYSE co-ruler Mars
- ✿ April 30: Mars 90-degrees square NYSE natal Moon
- ✿ April 30: Jupiter conjunct NYSE natal Saturn
- ✿ May 1: Jupiter conjunct NYSE natal Saturn
- ✿ May 3-4: FOMC meets
- ✿ May 5: Sun 90-degrees square NYSE natal Uranus
- ✿ May 7: Mars 90-degrees square NYSE natal Jupiter
- ✿ May 8: Mars 90-degrees square NYSE natal Mid-Heaven
- ✿ May 9: Moon at declination minimum
- ✿ May 9: Sun 120-degrees trine NYSE natal Mars
- ✿ May 14: Moon passes NYSE natal Mid-Heaven point
- ✿ May 24: Sun 90-degrees square NYSE natal Pluto
- ✿ May 14: Mars 90-degrees square natal Saturn
- ✿ May 14: Mercury retrograde ends
- ✿ May 15: Mercury at Aphelion

✧ May 16: Mars 90-degrees square natal Neptune.

✧ May 18: Sun conjunct NYSE natal Sun.

May - June 2023

Market action through late May and much of June 2023 will be influenced by the New Moon cycle commencing on May 19, 2023 (Sun at 27 Taurus) and running until June 18 (Sun at 26 Gemini). The lunation itself is not in harmful aspect to any other planets. In fact, it is 60-degrees from both Neptune and Mars, the NYSE co-ruling planets. However, Jupiter, Sun, Node, Moon, Uranus, and Mercury remain squeezed into a 30-degree portion of the zodiac. Compressed planetary energies can have consequences.

Key dates during this lunar cycle are:

✧ May 20: Moon passes 14 of Cancer. This is a week-end. Watch for a market reaction after the week-end.

✧ May 22: Moon at declination maximum

✧ May 24: Moon passes NYSE co-ruler Mars

✧ May 26: Moon VOC

✧ May 28: Venus completes its declination maximum

✧ May 29: Mercury at greatest westerly elongation

✧ May 29: Mars 90-degrees square NYSE natal Venus

✧ May 31: Moon VOC

✧ June 5: Moon at declination minimum

✧ June 9: Sun 90-degrees square NYSE natal Mars

✧ June 10: Moon passes NYSE natal Mid-Heaven

✧ June 12: Jupiter conjunct NYSE natal Venus

✧ June 14-15: FOMC meets.

- ☼ June 14: Sun 120-degrees trine NYSE natal Jupiter and Pluto. Sun 90-degrees square natal Mid-Heaven

- ☼ June 15: Mars conjunct NYSE natal Uranus.

June - July 2023

Market action through late June and much of July, 2023 will be influenced by the New Moon cycle that commences on June 18, 2023 (Sun at 26 Gemini) and runs until July 17 (Sun at 24 Cancer). The lunation itself is within 5 degrees of being 90-degrees square to the NYSE natal Mid Heaven. This square aspect implies some volatile excitability might be in store.

Key dates during this lunar cycle are:

- ☼ June 18: Moon at declination maximum

- ☼ June 19: Moon passes 14 of Cancer

- ☼ Sun 120-degrees NYSE natal Neptune

- ☼ June 21: Mars 120-degrees trine NYSE natal Moon

- ☼ June 22: Moon passes NYSE co-ruler Mars

- ☼ June 28: Mercury at Perihelion

- ☼ July 3: Moon at declination minimum

- ☼ July 4: Mars 120-degrees trine NYSE natal Saturn

- ☼ July 5: Mars 90-degrees square NYSE natal Sun

- ☼ July 5: Sun conjunct NYSE natal Ascendant

- ☼ July 6: Moon VOC

- ☼ July 7: Sun passes the NYSE natal Ascendant point

- ☼ July 8: Moon passes NYSE natal Mid-Heaven

- ☼ July 11: Sun 90-degrees square NYSE natal Moon

- ☼ July 9-17: Venus passes the same declination level as it was at in the 1792 NYSE natal horoscope
- ☼ July 15: Sun 90-degrees square NYSE natal Jupiter
- ☼ July 15: Sun 120-degrees trine NYSE natal Mid-Heaven
- ☼ July 16: Moon at maximum declination.

July - August 2023

Market action through late July and much of August 2023 will be influenced by the New Moon cycle that commences on July 17 (Sun at 24 Cancer) and runs until August 16 (Sun at 22 Leo). The lunation itself is at a 120-degree trine aspect to the NYSE natal Mid-Heaven point. Mercury and Venus events during this lunation will influence market trends.

Key dates during this lunar cycle are:

- ☼ July 19: Sun 90-degrees square NYSE natal Saturn and Neptune
- ☼ July 21: Moon passes NYSE co-ruler Mars
- ☼ July 23: Venus turns retrograde
- ☼ July 26-27: FOMC meets
- ☼ July 30: Moon at declination minimum
- ☼ August 3-14: Mars passes the same declination level as it was at in the 1792 NYSE natal horoscope
- ☼ August 4: Moon passes NYSE natal Mid-Heaven
- ☼ August 9: Mars conjunct NYSE natal Mars
- ☼ August 10: Mercury at greatest easterly elongation
- ☼ August 11: Mercury at Aphelion
- ☼ August 11: Sun 120-degrees trine NYSE natal Moon
- ☼ August 13: Venus at Inferior Conjunction

- ☼ August 13: Moon passes 14 of Cancer. Watch for an impact the next day (Monday)

- ☼ August 16: Moon VOC.

August - September 2023

Market action through late August and much of September 2023 will be influenced by the New Moon cycle that commences on August 16 (Sun at 22 Leo) and runs until September 15 (Sun at 21 Virgo). The lunation is not at any detrimental aspects to other planets. This month is highlighted by Venus continuing to be retrograde and by Mercury turning retrograde.

Key dates during this lunar cycle are:

- ☼ August 18: Moon passes NYSE co-ruler Mars

- ☼ August 19: Jupiter 90-degrees square NYSE natal Uranus through September 20

- ☼ August 19: Sun 120-degrees trine NYSE natal Saturn and square NYSE natal Sun

- ☼ August 23: Mercury turns retrograde

- ☼ August 23: Mars 120-degrees trine NYSE natal Sun

- ☼ August 27: Moon at declination minimum   September 1: Moon passes NYSE natal Mid-Heaven point

- ☼ September 2-October 7: Venus passes the same declination level as it was at in the 1792 NYSE natal horoscope

- ☼ September 3: Venus retrograde is complete

- ☼ September 8: Moon at declination maximum

- ☼ September 9: Moon passes 14 of Cancer

- ☼ September 11: Sun conjunct NYSE natal Mars

- ✪ September 14: Mercury retrograde completed
- ✪ September 15: Moon VOC.

September - October 2023

Market action through late September and much of October 2023 will be influenced by the New Moon cycle that commences on September 15 (Sun at 22 Virgo) and runs until October 14 (Sun at 20 Libra). This lunation is 180-degrees opposite to the NYSE natal Mid-Heaven point. The lunation is at the apex of an uncommon horoscope pattern called a *Kite* formation. In this case, the Kite is defined by Uranus, Neptune, Pluto and the lunation point. This formation suggests added market volatility.

Key dates during this lunar cycle are:

- ✪ September 16: Moon passes NYSE co-ruler Mars
- ✪ September 17: Mare 90-degrees square NYSE natal Ascendant
- ✪ September 20-21: FOMC meets
- ✪ September 20: Sun 120-degrees trine NYSE natal Sun
- ✪ September 22: Mercury at greatest westerly elongation
- ✪ September 23: Moon at declination minimum
- ✪ September 24: Mercury at Perihelion
- ✪ September 26: Moon VOC
- ✪ September 28: Moon passes NYSE co-ruler Neptune
- ✪ October 1: Mars conjunct NYSE natal Jupiter
- ✪ October 2: Mars 120-degrees trine NYSE natal Pluto
- ✪ October 6: Moon passes 14 of Cancer
- ✪ October 7: Sun 90-degrees square NYSE natal Ascenadant
- ✪ October 8: Mars conjunct NYSE natal Neptune.

October - November 2023

Market action through late October and much of November 2023 will be influenced by the New Moon cycle that commences on October 14 (Sun at 20 Libra) and runs until November 13 (Sun at 20 Scorpio). The lunation is within orb of being conjunct to Mars and 120-degrees trine to Saturn, all of which suggests heightened energy.

Key dates during this lunar cycle are:

- ☼ October 16: Sun conjunct NYSE natal Jupiter and trine natal Pluto
- ☼ October 16: Moon passes NYSE co-ruler Mars
- ☼ October 20: Moon at declination minimum
- ☼ October 26: Moon passes NYSE co-ruler Neptune
- ☼ November 1-2: FOMC meets
- ☼ November 1: Mars 120-degrees trine NYSE natal Ascendant
- ☼ November 2: Moon at declination maximum
- ☼ November 3: Moon passes 14 of Cancer
- ☼ November 3: Mars 90-degrees square NYSE natal Uranus
- ☼ November 6: Moon VOC
- ☼ November 6: Sun 120-degrees NYSE natal Ascendant
- ☼ November 7: Mercury at Aphelion
- ☼ November 15: Sun 120-degrees trine NYSE natal Mid-Heaven, Mars 90-degrees square NYSE natal Mid-Heaven
- ☼ November 15: Sun 90-degrees square NYSE natal Pluto
- ☼ November 16: Moon at declination minimum.

November - December 2023

Market action through late November and much of December 2023 will be influenced by the New Moon cycle that commences on November 13 (Sun at 20 Scorpio) and runs until December 13 (Sun at 19 Sagittarius). The lunation is conjunct Mars (NYSE co-ruler) and 120-degrees trine to the NYSE natal Mid-Heaven point. This is suggestive of an energized market. Sun and Mars completing their conjunction could add to the energy.

Key dates during this lunar cycle are:

- ☼ November 13: Moon at co-ruler Mars
- ☼ November 15: Mars 90-degrees square NYSE natal Pluto
- ☼ November 16: Moon at declination minimum
- ☼ November 22: Moon passes NYSE co-ruler Neptune
- ☼ November 29: Moon at declination maximum
- ☼ November 30: Moon passes 14 of Cancer
- ☼ December 4: Mercury at its greatest easterly elongation
- ☼ December 7: Sun 120-degrees trine NYSE natal Uranus
- ☼ December 10: Sun 90-degrees square NYSE natal Mars and trine NYSE natal Moon
- ☼ December 12: Moon passes NYSE co-ruler Mars
- ☼ December 13-14: FOMC meets.

December 2023 – January 2024

Market action through late December will be influenced by the New Moon cycle that commences on December 13 (Sun at 19 Sagittarius) and runs until January 11, 2024. The lunation is 90-degrees square the natal Mid-Heaven.

Key dates during this lunar cycle are:

- ✧ December 14: Moon at declination minimum
- ✧ December 15: Mars 120-degrees NYSE natal Uranus
- ✧ Sun 90-degrees square NYSE natal Mid-Heaven
- ✧ December 18: Sun is 120-degrees trine NYSE natal Saturn
- ✧ December 19: Moon passes NYSE co-ruler Neptune
- ✧ December 20: Mars 120-degrees trine NYSE natal Moon
- ✧ December 21: Mercury at Perihelion
- ✧ December 26: Mars 90-degrees square NYSE natal Mid-Heaven
- ✧ December 27: Moon at declination maximum
- ✧ December 27: Moon passes 14 of Cancer
- ✧ December 30: Mars 120-degrees trine NYSE natal Saturn.

CHAPTER FOURTEEN
Commodities 2023 Astrology

Gold

Investors who own Gold or related mining stocks are accustomed to routinely checking the price of Gold by tuning into a television business channel or perhaps obtaining a live online quote of the Gold futures price. What many do not realize is that quietly working behind the scenes to define the price of Gold is an archaic methodology called the *London Gold Fix.*

The 1919 Gold Fix Date

The London Gold Fix occurs at 10:30 a.m. and 3:00 p.m. local time each business day in London. Participants in the daily fixes are: Barclay's, HSBC, Scotia Mocatta (a division of Scotia Bank of Canada) and Societe Generale. These twice daily collaborations (some would say collusions) provide a benchmark price that is then used around the globe to settle and mark-to-market all the various Gold-related derivative contracts in existence.

The history of the Gold Fix dates back over one hundred years. On the 12th of September 1919, the Bank of England made arrangements with N.M. Rothschild & Sons for the formation of a Gold market in which there would be one official price for Gold quoted on any one day. At 11:00 a.m., the first Gold fixing took place, with the five principal gold bullion traders and refiners of the day present. These traders and refiners were N.M. Rothschild & Sons, Mocatta & Goldsmid, Pixley & Abell, Samuel Montagu & Co. and Sharps Wilkins.

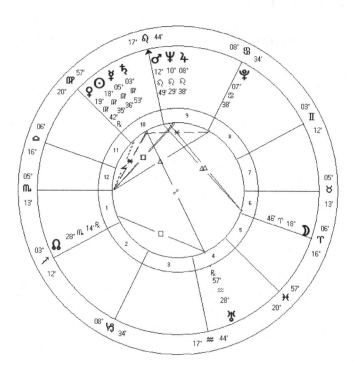

Figure 14-1
1919 London Gold Fix horoscope

The horoscope in Figure 14-1 depicts planetary positions at the 1919 Gold Fix creation date. Observations that jump off the page include: North Node had just changed signs, Venus was retrograde, Sun and Venus were conjunct, Mercury and Saturn were conjunct, Mars, Neptune and Jupiter were all conjunct at/near the Mid-Heaven point of the horoscope, and Saturn was 180-degrees opposite Uranus.

Gold investors who have been around for a while will remember the significant $800 per ounce price peak recorded by Gold in January 1980. To illustrate how astrology is linked to Gold prices, consider that at this price peak, the transiting North Node had just changed signs and was 90-degrees hard aspect to the natal Node in the 1919 horoscope. Consider too that Mars and Jupiter were both coming into a 0-degree conjunction with the natal Sun location in the 1919 horoscope.

As another price peak example, recall that Gold hit a significant peak in early September 2011 at just over $1900/ounce. At that peak, Sun and Venus were conjunct to one another as they were in the 1919 Gold Fix horoscope. What's more, they were within a few degrees of being conjunct to the natal Sun location in the 1919 horoscope.

In the few weeks that followed this 2011 peak, Gold prices plunged nearly $400 per ounce. But, then Gold found its legs again and began to rally. This rally seems directly related to Mars coming into a 0-degree conjunction to the Mars-Jupiter-Neptune location of the 1919 horoscope wheel.

In early August 2020, Gold again made a significant high, getting just above $2000/ounce. Sun conjunct to the 1919 Mars position along with Mars conjunct to the 1919 natal Moon are the obvious factors at work.

In March 2022, Gold again poked its head above the $2000/ounce mark. At the time, Sun was 180-degrees opposite the 1919 natal Sun, Jupiter was conjunct the 1919 natal Sun, and Mercury had just completed a conjunction with Saturn (these two are conjunct in the 1919 horoscope).

Such is the intriguing nature of Gold prices. I have studied past charts of Gold and am amazed at how many price inflection points are related in one way or another to the astrology of the 1919 Gold Fix horoscope wheel. To those readers who are of the opinion that Gold price is manipulated, your notion is indeed a valid one. I believe that planetary cycles are at the core of the secret language being spoken amongst those that play a hand in the price manipulation.

In January 2022, the Node changed signs as it was also making a 180-degree opposition aspect to the 1919 natal Node. Once the sign change was completed, Gold prices started to rally smartly. The month of May saw Venus passing the 1919 natal Moon location. May 16 saw an interim low at $1765/ounce. June had Mars passing the 1919 natal Moon location and as it did, Gold prices weakened further. August saw Venus passing the 1919 natal Jupiter, natal Neptune and natal Mars points. As it did, Gold prices peaked and started heading lower. September saw Venus passing the 1919 natal Sun location.

- ☼ **In 2023, the Node will change signs and move into Aries in mid-July.**

- ☼ **Venus will pass the 1919 natal Moon location in early March.**

- ☼ **Mars will not transit past the 1919 natal Moon location in 2023.**

- ☼ **Venus will pass the 1919 natal Jupiter, natal Neptune and natal Mars points in June, 2023.**

- ☼ **Mars will pass the 1919 natal Sun location in August.**

- ☼ **Venus will pass the natal Sun location in October.**

Declination

On September 12, 1919 when the Gold Fix mechanism was started, Mars was at -1.5 degrees declination, Venus was also at -1.5 degrees, Moon was at 0-degrees declination and Sun was at +4 degrees declination. If four celestial bodies all at or near 0-degrees declination seems like a co-incidence, it is not. This date in 1912 was picked for these obvious planetary features.

Examination of a Gold price chart for 2022 shows that these declination levels do align to short term trend changes.

In 2022, Mars was at or near -1.5 degrees declination around May 23-28. Gold did not move much higher in price after this declination event.

Venus was at or near -1.5 degrees declination around April 30-May 4 and again October 1-6. In early May, Gold was trending strongly to the downside and was unaffected by the declination event.

In 2022, Sun was at or near 4 degrees declination around March 30 and again around September 12. A look at a Gold chart shows that March 29 had a significant intraday price swing of $41/ounce.

> ☼ **In 2023, Sun will be at 4 degrees declination around March 30 and again around September 12. Venus will be at or near -1.5 degrees declination February 17-20 and again November 12-16. Mars will be at -1.5 degrees declination September 2-8.**

Outer Planet Transits of the 1919 Natal Points

> ☼ **In mid-March, 2023 Jupiter will move past the 1919 Gold Fix natal Moon point.**

1974 Gold Futures Date

Gold futures contracts started trading in North America on the New York Mercantile Exchange on December 31, 1974. Figure 14-2 illustrates the planetary positions in 1974 at the first trade date of Gold futures.

Note that in the 1919 chart, Mars and Neptune are conjunct one another. In the 1974 chart, Mars and Neptune are also conjunct one another.

Next, consider why the New York Mercantile Exchange would launch a new futures contract on December 31, a time when most staff would be off for Christmas holidays. If this seems more than a bit odd, you are not alone in your thinking.

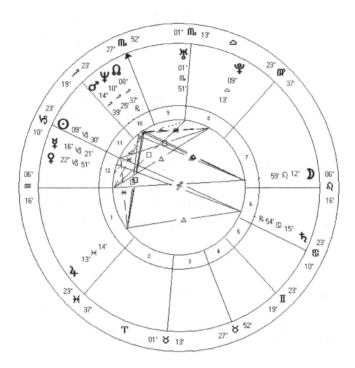

Figure 14-2
Gold futures First Trade horoscope

Note the location of Moon in the 1974 horoscope at 12 degrees of Leo. Now, look at the 1919 horoscope and observe that 12 degrees of Leo is where Mars and Neptune are located. The 1919 Gold Fix Ascendant is located 90-degrees square to the 1974 Futures Ascendant if one assumes a first trade start time of 9:00 a.m. which is a more than reasonable assumption. I take these curious placements as further evidence of a deliberately timed astrological connection between Gold price, the 1919 Gold Fix date, and the 1974 first trade date for Gold futures.

Venus and Mars transits

Times when Venus and Mars transit past key points in the 1974 first trade horoscope deserve attention:

In early January 2022, Mars passed conjunct to the 1974 natal Mars location. A $50/ounce rally was the result.

In early February 2022, transiting Mars passed the 1974 natal Sun location. Gold prices rallied from $1800 to well over $2000 per ounce. This rally was aided by Venus emerging from retrograde in late January to be conjunct the 1974 natal Sun.

In mid-April, Venus passed the 1974 natal Jupiter point. Starting on April 18, Gold prices began a decline that shaved off over $200 from Gold prices.

In early May as Mars passed the natal Jupiter point, Gold prices reacted in a volatile manner, swinging around in an $80 range as this transit unfolded.

In late July/early August 2022, Venus passed the 1974 natal Saturn point. A counter-trend rally that was unfolding suddenly lost momentum and the price of Gold rolled over.

In late-August, Venus passed the natal Moon point. What appeared to be the humble start of a counter-trend rally failed at this transit and Gold resumed its downtrend.

In early October as Venus was set to pass the natal Pluto point, Gold surrendered another of its brief counter-trend rallies and move further to the downside.

- ☼ **For 2023, Venus will pass the 1974 natal Saturn point in late May. Venus will pass the natal Jupiter point in mid-February. Venus will pass the natal Moon location in mid-June.**

- ☼ **For 2023, Mars will pass the natal Saturn point in mid-April. Mars will pass the natal Moon point in mid-June.**

Mercury Retrograde

Another valuable tool for Gold traders to consider is Mercury retrograde events. Watch for technical chart trend indicators to suggest a short term trend change at a retrograde event.

The chart in Figure 14-3 illustrates the connection between the three Mercury phenomena and Gold prices in 2022. Each retrograde event aligned to a V-bottom pattern.

For 2023, Mercury will be:

- ☼ **retrograde from December 29, 2022 through February 17**

- ☼ **retrograde from April 21 through May 14**

- ☼ **retrograde from August 23 through September 14**

- ☼ **retrograde from December 13 through January 1, 2024.**

Figure 14-3
Mercury retrograde and Gold prices

The planetary declination intrigue noted with the Gold Fix historical date extends also to the 1974 Gold futures date. At December 31, 1974,

Mars and Venus were at -22 degrees declination and Sun was at -23 degrees declination. It surely was not a co-incidence to select a first trade date with three celestial bodies all at the same declination.

For 2022, Mars was at or near its minimum declination in late January. This event supported a price rally. Venus was at or near -23 degrees declination in mid-December, 2022.

☼ **For 2023, Mars will be at its minimum declination in December. Venus will be near a declination of -22 degrees in early January.**

Silver

Silver futures started trading on a recognized financial exchange on July 5, 1933. Figure 14-4 shows the First Trade horoscope for Silver futures in geocentric format.

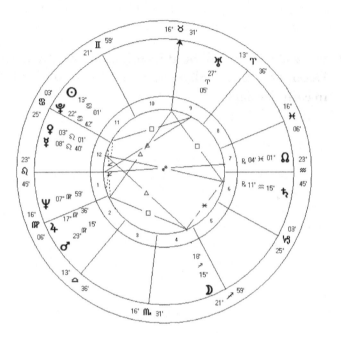

Figure 14-4
Silver futures First Trade horoscope

I am intrigued with this first trade date. Silver could have started trading anytime in 1933. July 4, 1776 is a critical date in US history and on this date Sun was at 14 Cancer. Recall that 14 Cancer also figures prominently in the first trade horoscope of the NYSE. In 1933, markets would have been closed for the 4th of July celebrations. A first trade date of July 5, is as close as authorities could come to the July 4 date. On July 5, Sun at 13 degrees is within a degree of the critical 14 of Cancer point.

My research has shown that the times when transiting Sun, transiting

Mars and transiting Jupiter make hard aspects to the 1933 natal Sun point at 12 degrees Cancer should be watched carefully for evidence of trend changes and price inflection points.

Natal Sun Aspects

Figure 14-5 illustrates several 2022 conjunct and square aspects to the natal Sun location.

Figure 14-5
Silver futures and natal Sun

In March 2022, Sun making a square aspect to the 1933 natal Sun resulted in a bottoming chart formation. Days after this transit, the downtrend resumed. Mars making a 90-degree hard aspect to natal Sun in June brought a halt to a counter-trend rally. The Sun conjunct natal Sun aspect in early July confirmed the downtrend. Mars making a 90-degree hard aspect to natal Sun in October produced a sharp swing high reversal. Jupiter made a square aspect to the 1933 natal Sun in July and August. As this aspect ended at the end of August, a significant low was recorded. Whether this low holds going forward remains to be seen.

☼ **In 2023, Mars will pass the 1933 natal Sun location in mid-April and will make a 90-degree aspect in mid-September. Sun will aspect the natal Sun location in early April (90-degrees), early July (0-degrees), and early October (90-degrees). In late February through early March Jupiter will pass 90-degrees to the natal Sun location.**

Declination

Planetary declinations should also be considered when studying price action of Silver futures. In particular, the declination of Venus, Sun, Moon, and Mars deserve watching. Venus had just made its maximum declination in 1933 as Silver futures were starting to trade for the very first time. Sun was at 22-degrees declination. Moon was at a declination minimum and Mars was at 0-degrees declination.

As discussed earlier in Chapter 1, the Sun is not moving up and down in its declination. Rather, as planet Earth orbits the Sun, Earth exhibits a change in the tilt relative to its axis. This affects the angle of the Sun's rays striking the Earth. Astronomers have thus adopted the standard approach of referring to the declination of the Sun, not to the declination of the Earth.

At the Sun's declination maximum on June 21, 2022, Silver began a precipitous move lower. At the maximum declination of Venus in mid-July 2022, Silver prices made a bottoming pattern and a brief rally unfolded. As Mars passed through 0-degrees of declination in late May 2022, the price of Silver topped and began to decline. The chart in Figure 14-6 illustrates the alignment between Moon declination minima and price inflection points on Silver price.

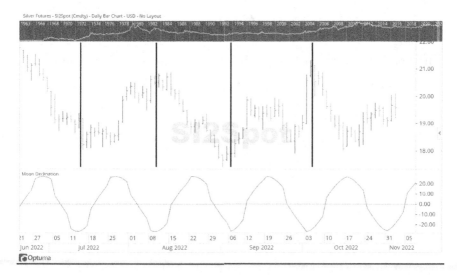

Figure 14-6
Silver futures and natal Sun

☼ **For 2023, Sun will be at its maximum declination at the Summer Solstice on June 21. Sun will at its minimum declination at the Winter Solstice on December 21.**

☼ **For 2023, Venus will exhibit its maximum declination either side of May 10.**

☼ **For 2023, Mars will be at 0-degrees declination for several days either side of November 10.**

☼ **For 2023, Moon will be at its minimum declination: January 20, February 16, March 15, April 12, May 9, June 5, July 2, July 30, August 27, September 24, October 20, November 16, and December 14.**

Copper

The first trade date for Copper futures was July 29, 1988. Figure 14-7 illustrates the first trade horoscope.

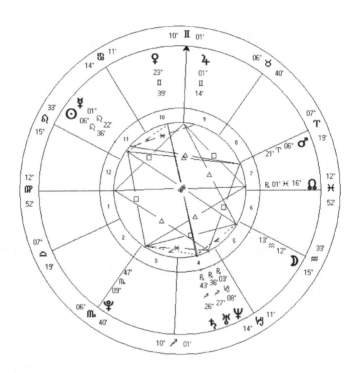

Figure 14-7
Copper futures first trade horoscope

Inferior Conjunction

This horoscope wheel features an Inferior Conjunction of Mercury in the sign of Leo. Also, notice in this horoscope that the first trade date is that of a Full Moon.

An Inferior Conjunction of Mercury marks the start of a new Mercury cycle around the Sun. Mercury Inferior Conjunction events always occur in association with Mercury being retrograde.

Mercury Retrograde

Figure 14-8 shows Copper prices overlaid with Mercury retrograde events. Knowing that retrograde events are approaching, one should watch for short term trend changes using a suitable trend indicator.

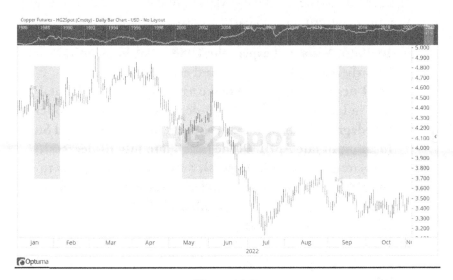

Figure 14-8
Copper and Mercury retrograde events

The retrograde event in early 2022 caused some added price volatility, but the price uptrend remained intact. The retrograde event in May 2022 was unique in that price recorded a swing low at the start of retrograde and a swing high at the end of retrograde. The September retrograde event seems to have established an interim level of support.

For 2023, Mercury will be:

☿ **retrograde from December 29, 2022 through February 17**

☿ **retrograde from April 21 through May 14**

☿ **retrograde from August 23 through September 14**

☿ **retrograde from December 13 through January 1, 2024.**

Copper prices are also influenced by Sun and Mars making aspects to the 1988 natal Sun and natal Mars points.

In 2022, Sun made a 90-degree square aspects to natal Sun in late April. This aligned to a steep drawdown in Copper prices. Mars made a 90-degree square to the 1988 natal Mars in mid-July, 2022. This aligned to a V-bottom pattern on the chart.

☼ **In 2023, Mars will aspect the 1988 natal Mars point in early April (90-degrees), and early September (180-degrees). Mars will make aspects to the 1988 natal Sun in late May-early June (0-degrees) and mid-October (90-degrees). Sun will make aspects to the 1988 natal Sun location in late April (90-degrees), late July (0-degrees), and late October (90-degrees). Sun will make aspects to the 1988 natal Mars point in late March (0-degrees), late June (90-degrees), late September (180-degrees), and late December (90-degrees).**

Declination

At the 1988 first trade date, Mars was at -1.4 degrees of declination. Mercury was at or near its maximum declination and Moon was at its minimum declination.

In 2022, Mercury was at or near maximum declination around May 5 and again around July 11. Days after the May 5 event, Copper prices recorded a swing low and began to move higher. Mars passed through -1.5 degree level of declination in early June as Copper prices started to sell off in a significant way.

☼ **In 2023, Mars will be at its 1988 natal declination level between November 9 and 19. Mercury will be at its maximum declination for several days either side of July 1.**

Canadian Dollar, British Pound, and Japanese Yen

These three currency futures all started trading on May 16th, 1972 at the Chicago Mercantile Exchange. The horoscope in Figure 14-9 illustrates planetary placements at this date. It is interesting to note that Mars is 180-degrees opposite Jupiter. This suggests that Mars and Jupiter may play a role in price fluctuations on these currencies. Mars is also 0-degrees conjunct to Venus, suggesting another cyclical relationship.

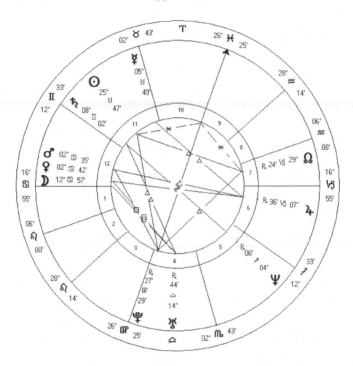

Figure 14-9
Pound, Yen, Canadian First Trade horoscope

Outer Planet Aspects

☿ **Jupiter will start 2023 in a square position to natal Mars and natal Venus. By early March, Jupiter will have moved past the natal Moon location.**

145

☼ **In early February, Jupiter will be 60-degrees (sextile) to the natal Saturn location.**

☼ **In January-February, 2023 Saturn will be at a square aspect to the natal Sun location.**

☼ **Through March, April, and May Saturn will be at a square aspect to the natal Saturn point.**

The Transiting Mars-Venus Influence

In 2022, February and March delivered a Mars/Venus conjunction pattern. The Canadian dollar rallied 2.5 cents. The British Pound traded in a sideways pattern, taking a break from the prevailing downtrend that had started in 2021. The Japanese Yen started on a perilous downward journey.

☼ **In 2023, Mars and Venus will be close to conjunction in June and part of July.**

Natal Mars and Natal Sun transits

Transiting Mars passing natal Sun, natal Mars and natal Moon are events that British Pound currency traders may also wish to focus on.

To illustrate, the chart in Figure 14-10 shows that in mid-2021 at a Mars/natal Moon aspect, the Pound hit a high of 1.42 versus the US Dollar. In August 2022, Mars conjunct the 1972 natal Sun accelerated the decline of the Pound to lows not see in twenty years.

For 2023, transiting Mars will make conjunct aspects to the natal Mars, and Moon points from March 23 through April 23.

Figure 14-10
British Pound Mars passing natal Mars, Moon, and Sun

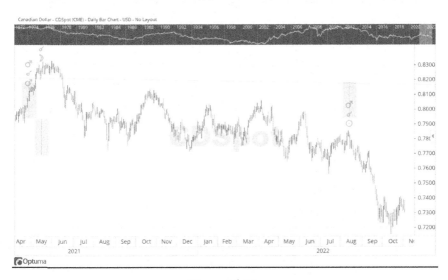

Figure 14-11
Canadian Dollar aspects to natal Sun and Mars

The chart in Figure 14-11 illustrates the effect on the Canadian Dollar
of transiting Mars making 0-degree aspects to the 1972 natal Sun, natal

Mars and natal Moon points. The pattern is very similar to that of the British Pound.

Mercury Retrograde

Currency traders should pay close attention to Mercury retrograde events as they can bear a good alignment to trend changes on the Pound, Yen, and Canadian Dollar.

The British Pound price chart in Figure 14-12 has been overlaid with Mercury retrograde events. The correlation to swing highs and lows is plainly evident. Perhaps new Prime Minister Liz Truss should have waited until Mercury retrograde was over before announcing fiscal and monetary strategy.

Figure 14-12
British Pound and Mercury retrograde

For 2023, Mercury will be:

☼ **retrograde from December 29, 2022 through February 17**

☼ **retrograde from April 21 through May 14**

☼ **retrograde from August 23 through September 14**

☼ **retrograde from December 13 through January 1, 2024.**

Declination

At the 1972 first trade date, Mars, Venus and Moon were all at or near their declination maxima. A look at past price performance suggests Mars and/or Venus at or near declination maxima align to trend changes.

For 2022, Venus exhibited its maximum declination several days on either side of July 23. Mars was at or near its maxima in early December.

☼ **For 2023, Venus will be at its declination maximum for several days either side of May 9. Mars will spend the first three months of the year at its declination maximum.**

Euro Currency

The Euro became the official currency for the European Union on January 1, 2002 when Euro bank notes became freely and widely circulated.

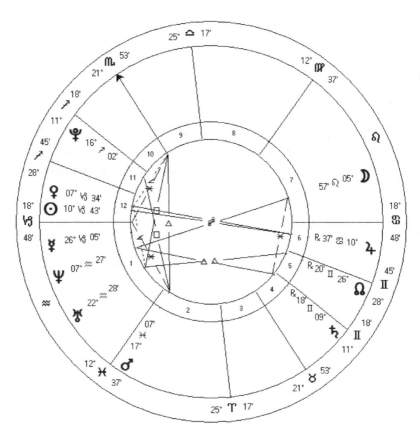

Figure 14-13
Euro Currency First Trade horoscope

Outer Planet Natal Aspects

☼ **From January through late February, 2023 Saturn will be conjunct to natal Uranus.**

☼ **From late February through early April, Jupiter will be conjunct the natal Jupiter location.**

Sun Transits of Natal Sun

Events of transiting Sun and Mars making 0, and 90-degree aspects to the natal Sun position in the Euro 2002 First Trade horoscope are worth watching as they often align to inflection points on the Euro. In 2022 notice how these transits all came within close proximity to price inflection points.

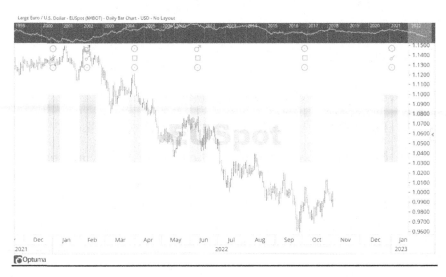

Figure 14-14
Natal Transits and the Euro Currency

For 2023, transiting Sun will be:

- ☼ **At a 0-degree conjunction to natal Sun during the first week of January**

- ☼ **90-degrees to natal Sun March 26 through April 7**

- ☼ **90-degrees to natal Sun from September 28 through October 9.**

For 2023, transiting Mars will be at a 90-degree aspect to natal Sun in mid-September.

Declination

At this 2002 first trade date, Venus and Sun were at or near their declination minima. Moon was at its declination maxima. Mars was at -5 degrees declination.

In 2021, Venus was at its declination low in early November. The Euro responded by falling from the 1.17 level to near 1.12.

In May 2022 as Mars passed -5 degrees of declination, the Euro staged a brief rally.

Venus will be at its declination low in mid-December 2022.

- ☼ **For 2023, Venus will not exhibit a declination low.**

- ☼ **Mars will pass -5 degrees declination in mid-September.**

- ☼ **For 2023, Moon will be at its minimum declination: January 20, February 16, March 15, April 12, May 9, June 5, July 2, July 30, August 27, September 24, October 20, November 16, and December 14.**

Australian Dollar

Australian dollar futures started trading on the Chicago Mercantile Exchange on January 13, 1987. As the horoscope in Figure 14-15 shows, Sun and Mercury are at Superior Conjunction at 22-23 degrees Capricorn.

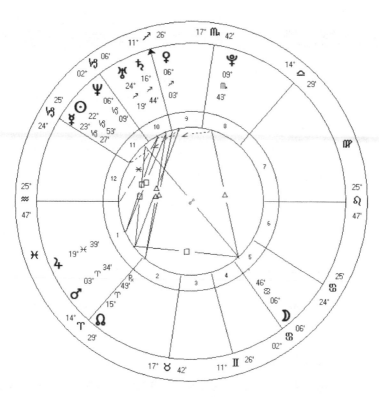

Figure 14-15
First trade horoscope of Australian Dollar futures

Outer Planet Aspects

☼ **In early January, 2023 Jupiter will pass conjunct to the natal Mars point.**

☼ **In mid-March, Jupiter will pass conjunct to the natal Node point.**

Mercury Retrograde

Times when Mercury is retrograde should be considered when trading Australian Dollar futures. The chart in Figure 14-16 has been overlaid with Mercury retrograde events. These events often align to price inflection points.

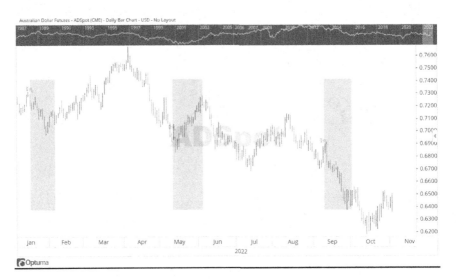

Figure 14-16
Australian Dollar and Mercury Conjunctions

For 2023, Mercury will be:

- ☿ **retrograde from December 29, 2022 through February 17**

- ☿ **retrograde from April 21 through May 14**

- ☿ **retrograde from August 23 through September 14**

- ☿ **retrograde from December 13 through January 1, 2024.**

Declination

At this 1987 first trade date, Mercury was at or near its declination low. Moon was at its maximum declination. Mars was at or near 0-degrees of declination. Venus was at -18 degrees of declination.

In 2022, Mars passed through 0-degrees of declination in late May. This passage halted a rally and set the Australian dollar on a steep downward trajectory. Mercury will be at its low in early December. Venus was be at -18 degrees of declination in November, 2022.

- ✷ **In the second week of January, 2023 Venus will pass -18 degrees declination.**

- ✷ **Mars will pass through zero degrees declination low in late August.**

- ✷ **Moon will be at its maximum declination: January 6, February 2, March 2, March 28, April 25, May 22, June 18, July 15, August 12, September 8, October 5, November 2, November 29, and December 26.**

30-Year Bond Futures

30-Year Bond futures started trading in Chicago on August 22, 1977. Figure 14-17 presents the first trade horoscope for this date.

Figure 14-17
First trade horoscope for 30-Year Bond futures

Outer Planet Aspects

☼ **In January 2023, Jupiter will be square to the natal Jupiter point.**

Mars Transits

Money and its cost (ie Bond yields) are critical factors to the functioning of an economy. It interesting that the 30-Year Bond natal horoscope has Mars at 24 Gemini. In the 1776 natal horoscope of the USA, Mars just so happens to be at 21 Gemini. Was this a factor in selecting the first trade date of the 30 Year Bond futures? Had the first trade date been set at the previous Friday (Aug 19), Mars would have been exactly at 21 of Gemini.

My research has indicated that events of transiting Mars making 0 and 90-degree aspects to the natal Mars position at 24 degrees Gemini are valuable tools for interpreting Bond prices. Also, Mars making aspects to natal Sun at 29 Leo is important to watch.

Figure 14-18 illustrates Bond price performance with the Mars/natal Mars and Mars/natal Sun transits overlaid. Weakness in the Bond market manifested in December 2021 when the Federal Reserve announced an end to its liquidity injections. This policy shift came precisely as Mars was making a square aspect to the 1977 natal Sun. In May 2022, a Mars 90-degree aspect to natal Mars ended a brief counter-trend rally.

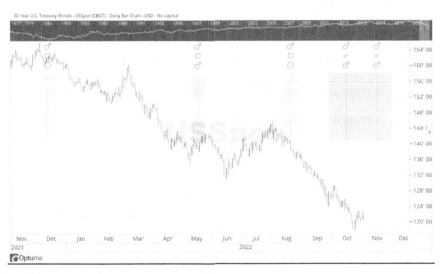

Figure 14-18
30-Year Bonds and Mars in aspect to natal Sun and Mars

Between August 10 and 28, transiting Mars made a 90-degree aspect to natal Sun. Mere days prior to this aspect starting, the price of 30-year Bonds began another journey lower.

For 2023:

☼ **Mars will be conjunct natal Mars in the first 3 weeks of March**

☼ **Mars will be conjunct natal Sun in the first 2 weeks of July**

Mercury Retrograde

In the first trade horoscope in Figure 14-17 note that the position of Mercury (at 20 Virgo) is further delineated by a letter *S*. This letter denotes *stationary*. The term stationary refers to the day immediately prior to a planet turning retrograde and starting to move backwards in the zodiac wheel. This first trade date of August 22, 1977 comes one day prior to Mercury turning retrograde. Was this also a factor in selecting this first trade date?

For 2023, Mercury will be:

☼ **retrograde from December 29, 2022 through February 17**

☼ **retrograde from April 21 through May 14**

☼ **retrograde from August 23 through September 14**

☼ **retrograde from December 13 through January 1, 2024.**

Declination

At the 1977 first trade date, Mars was at its maximum declination and Mercury was at 0-degrees declination. Venus was at 20-degrees declination.

Mars at its declination maximum in May 2021 saw a price rally get underway which took Bond prices from the 155 level to the 167 level.

In late March 2022, Mercury at 0-degrees of declination saw the failure of a brief rally attempt and a move lower in Bond prices. Mercury at 0-degrees declination again in late September saw a brief attempt at a rally which soon faded. At this time of writing, Mercury is once more passing 0-degrees declination.

Venus at maximum declination of 22 degrees in late July 2022 was very close to its natal declination point. What ensued on the 30-Year Bond market was a serious sell-off.

At this time of writing, Mars at maximum declination is still 6 weeks away and will occur in early December. Mars will hover at or near maximum declination through the end of 2022.

- ☼ **In 2023, Mars will be at its maximum declination for the first three months of the year.**

- ☼ **In 2023, Mercury will pass through 0-degrees declination in early March, late August, and early October.**

- ☼ **In early April and again in mid-June, 2023 Venus will pass through 20-degrees declination.**

10-Year Treasury Note Futures

10-Year Treasury Notes started trading in Chicago on May 3, 1982. Figure 14-19 presents the first trade horoscope for this date.

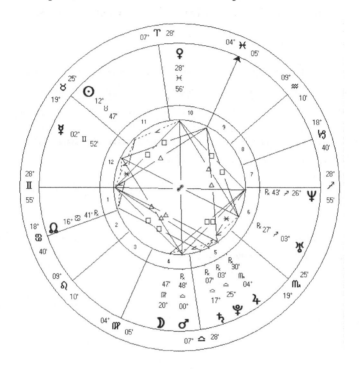

Figure 14-19
First trade horoscope for 10-Year Treasury Notes

Outer Plant Aspects

☼ **In mid-March, Jupiter will be square the Node.**

☼ **In early August, Jupiter will be conjunct the natal Sun point.**

Retrograde

Note that in this first trade horoscope Mars is denoted *Rx* which stands for *retrograde*. Therein rests another valuable clue. Looking back at past Treasury Note performance, significant price inflections at Mars retrograde events can be seen. Mars was retrograde from early September through to mid-November 2020. This event turned out to be a price high on the 10-Year Treasury market.

In 2022, Mars turned retrograde on October 30. What was looking like a counter-trend rally was quickly quelled as retrograde set in. The 10-Year Treasuries then weakened further, adding upward pressure to interest rates.

Mars retrograde will continue until January 11, 2023.

Mercury retrograde events also bear watching when following price action on the 10-Year Treasury Notes.

For 2023, Mercury will be:

- ☿ **retrograde from December 29, 2022 through February 17**
- ☿ **retrograde from April 21 through May 14**
- ☿ **retrograde from August 23 through September 14**
- ☿ **retrograde from December 13 through January 1, 2024.**

Declination

At the 1982 first trade date, Mars and Venus were within a degree of each being at 0-degrees declination. An examination of past price data reveals a solid alignment to these planets at 0 declination and short-term trend swings.

In 2022, Mars passed through 0-degrees declination on June 1. Two days prior to this event, 10-Year Treasuries resumed their downtrend, pushing interest rates higher.

In 2022, Venus passed 0 declination around May 6 and again around October 1. The May 6 date marked the start of a brief countertrend rally.

For 2023:

☼ Venus will pass through 0 declination in late February and again in mid-November

☼ Mars will pass through 0 declination at the end of August

Wheat, Corn, and Oats

1877 Futures

Wheat, Corn and Oats futures all share the same first trade date of January 2, 1877. The horoscope in Figure 14-20 shows planetary placements at that date. Note how the 14 Cancer point appears exactly opposite the Ascendant point. The choice of this first trade date is is not a random occurrence.

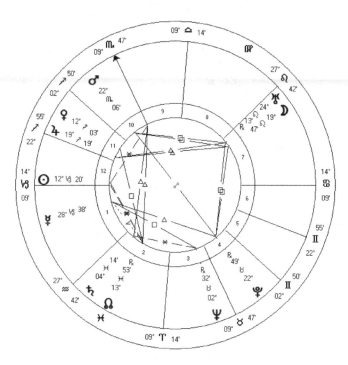

Figure 14-20
First trade horoscope for Wheat, Corn and Oats futures

CBOT 1848

W.D. Gann was also known to follow a first trade horoscope wheel from April 3, 1848, the date the Chicago Board of Trade (CBOT) was founded. Figure 14-21 shows this horoscope wheel.

In the 1877 Wheat/Corn natal horoscope, the Sun is at 12 Capricorn, exactly square to the location of Sun in the 1848 horoscope.

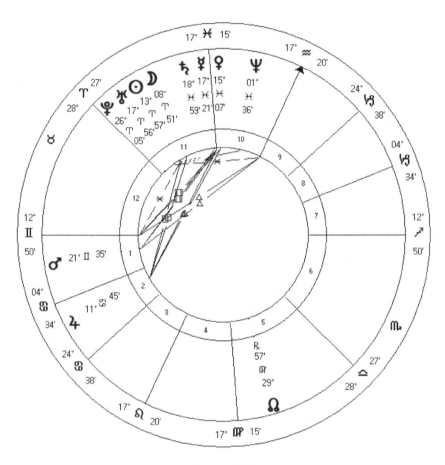

Figure 14-21
1848 first trade horoscope for CBOT

Sun, Jupiter, and Mars

Events of transiting Sun making 0, and 90-degree aspects to the natal Sun position in the 1877 first trade horoscope or the 1848 CBOT natal horoscope can be used as a tool to guide traders through the shorter term price volatility of Wheat, Corn and Oats. Because of the peculiar alignment of these two horoscopes, a 0-degree conjunction to natal Sun in the 1877 horoscope will be a 90-degree square to the natal Sun of the 1848 horoscope.

As Figure 14-22 shows, in early January 2022, Sun passing conjunct to the 1877 natal Sun location triggered a rally that took Wheat prices up $6 per bushel. A square event in late March aligned to a V-bottom chart formation. At this time of writing, a square aspect seems to be causing some short term weakness.

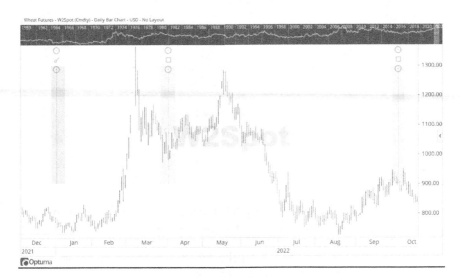

Figure 14-22
Wheat futures and Sun/natal Sun aspects

Mars is also a planet to watch closely. As Figure 14-23 illustrates, Mars making a conjunction to the 1877 natal Sun in early February 2022 added impetus to a rally in Corn (and Wheat) prices. Mars making a square aspect to the 1877 natal Sun in June, 2022 caused an increase in Corn prices. As this aspect faded, Corn prices reversed course. This same aspect caused Wheat to trade in a sideways consolidation pattern. As the aspect faded, Wheat prices began to tumble.

Figure 14-23
Corn futures and Mars/natal Sun aspects

In 2023:

☼ **Mars will be at a 90-degree aspect to the 1877 natal Sun in mid-September. from February 5 to 16**

☼ **In 2023, Sun will aspect the 1877 natal Sun on January 2 (0-degrees), late March (90-degrees), late June (180-degrees), and late September (90-degrees)**

Declination

A deeper examination of declination reveals that in 1848 when the Chicago Board of Trade was founded, Mars was at its maximum declination and Moon was at 0-degrees declination. In 1877, when Wheat, Corn and Oats started trading, Venus was within about 3 degrees of its minimum declination.

In 2022, Venus was at its declination minimum in mid-December. Mars did not record a maximum declination.

In 2023:

☼ **Mars will be at maximum declination for the first three months of the year**

☼ **Venus will not record a minimum declination**

Mercury Retrograde

Mercury retrograde plays a role in price pivot points on the grains. The price chart of Wheat futures in Figure 14-24 has been overlaid with Mercury retrograde events. In early 2022, a retrograde event helped support the start of a move higher in price. In May 2022, a retrograde event helped turn the trend negative. The retrograde event in September 2022 seems to have put a damper on further price increases in the near term.

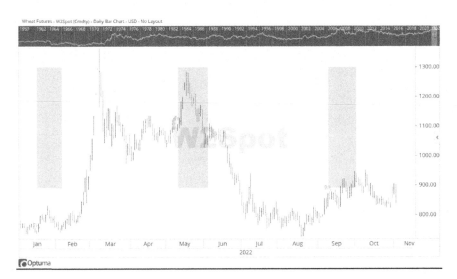

Figure 14-24
Wheat prices and Mercury retrograde events

For 2023, Mercury will be:

- ☼ retrograde from December 29, 2022 through February 17
- ☼ retrograde from April 21 through May 14
- ☼ retrograde from August 23 through September 14
- ☼ retrograde from December 13 through January 1, 2024.

Soybeans

Soybean futures started trading in Chicago on October 5, 1936. The horoscope in Figure 14-25 illustrates the planetary placements at that time. What is intriguing is the location of the Sun. Notice how it is exactly 90-degrees to the location of the Sun in the first trade horoscope for Wheat, Corn and Oats. Notice Sun is 180-degrees from the Sun in the 1848 CBOT natal chart. As I have previously suggested, the regulatory officials who determined these first trade dates knew more about astrology than we may think. Moreover, if one makes the reasonable assumption of a 7:00 a.m. first trade, the Mid Heaven (MH) is at 14 Cancer. Studying the 1877 natal horoscope for Corn and Wheat reveals a further similarity to the 1936 Soybeans horoscope. Jupiter is at 18-19 degrees Sagittarius in both horoscopes.

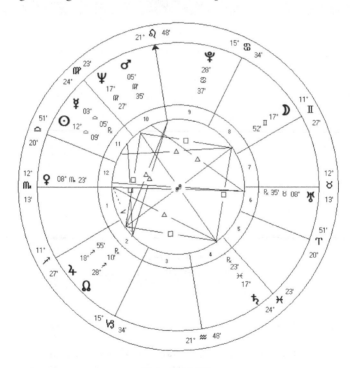

Figure 14-25
Soybeans first trade horoscope

Natal Transits

Events of transiting Sun and Mars making 0, and 90-degree aspects to the natal Sun position in the 1936 first trade horoscope can be used to navigate the volatility of the Soybean market.

Figure 14-26 illustrates the effect of transiting Sun and Mars making 0 and 90-degree aspects to the natal Sun position. In February 2022, a price high (and reversal) came days after a Mars/natal Sun square aspect. In late June 2022, a square aspect of Sun to natal Sun triggered a severe price drawdown. At this time of writing, Sun has just made a conjunction to the natal Sun location.

Figure 14-26
Soybeans and natal Sun aspects

In 2023:

- ☼ **Sun will aspect the 1936 natal Sun point in early January (90-degrees), early July (90-degrees), and early October (0-degrees)**

- ☼ **Mars will aspect the 1936 natal Sun in mid-April (90-degrees) and in mid-September (0-degrees)**

Mercury Retrograde

Mercury retrograde events also contribute to the price behavior of Soybeans. The Soybeans chart in Figure 14-27 illustrates the Mercury retrograde effect. The retrograde event in early 2022 came amid an up-trending market and did not disturb the trend. A price drawdown in April was interrupted right at the start of retrograde. The September event aligned to an effort to retrace a price gap. Once the gap was filled, the price of Soybeans again turned lower.

Figure 14-27
Soybeans and Mercury retrograde

For 2023, Mercury will be:

☿ **retrograde from December 29, 2022 through February 17**

☿ **retrograde from April 21 through May 14**

☿ **retrograde from August 23 through September 14**

☿ **retrograde from December 13 through January 1, 2024.**

Declination

Recall that the founding of the Chicago Board of Trade in 1848 had Mars at its maximum declination and Venus near its declination low.

In November 2021, Soybeans were trading at a low level of $11.81/bushel. Venus making its declination minimum played a role in starting a rally that took prices to over $17/bushel. Venus will again be at its declination low in early 2024.

In late April-early May 2021, Mars was at its declination maximum. Soybeans recorded an interim high at just of $16/bushel and then began to retreat.

In 2023:

✪ **Mars will be at maximum declination for the first 3 months of 2023.**

✪ **Venus will not record a declination minimum.**

Crude Oil

West Texas Intermediate Crude Oil futures started trading for the first
time in New York on March 30, 1983. A unique alignment of celestial
points can be seen in the horoscope in Figure 14-28. Notice how Mars,
North Node, (Saturn/Pluto/Moon), and Neptune conspire to form a
rectangle.

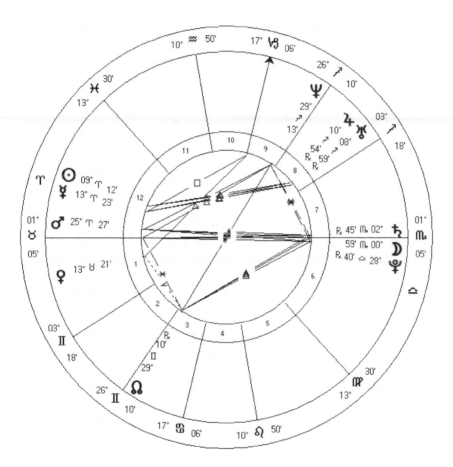

Figure 14-28
Crude Oil first trade horoscope

Transits of the Natal Rectangle

Crude Oil can be a complex instrument to analyze using astrology. Given the peculiar rectangular shape that appears in the horoscope, one strategy for analyzing Crude Oil is to use natal transits with a focus on transiting Sun and transiting Mars making 0-degree aspects to the four corner points of the rectangle.

The tenor of the price reaction as Sun and Mars each pass the various corner points varies from year to year. Nevertheless, this strategy remains viable.

For 2022, the four corners of the peculiar rectangle were passed by as follows:

- ☼ Sun transited 0-degrees to the natal Mars location from April 9 through April 22. Crude prices moved from $94 to $109 per barrel.

- ☼ Sun transited 0-degrees to the natal Node location from June 7 through June 22. The trend on Oil rolled over and turned negative.

- ☼ Sun transited 0-degrees to the natal (Saturn/Pluto/Moon) location from October 18 through November 1.

- ☼ Sun transited 0-degrees to the natal Neptune location from December 14 through the end of the year

- ☼ Mars transited past the natal Mars location between June 22 and July 5. Oil broke beneath $100 on July 5 reinforcing the downtrend.

In 2023, the four corners of the peculiar rectangle will be passed by as follows:

- ☼ **Sun will transit 0-degrees to the natal Mars location from April 9 through April 22.**

- ☼ **Sun transited 0-degrees to the natal Node location from June 7 through June 22.**

- ☼ **Sun will transit 0-degrees to the natal (Saturn/Pluto/Moon) location from October 18 through November 1.**

- ☼ **Sun will transit 0-degrees to the natal Neptune location from December 14 through the end of the year.**

- ☼ **Mars will transit past the natal Node location from March 19 to April 2.**

- ☼ **Mars will transit 0-degrees to the natal (Saturn/Pluto/ Moon) location from October 10 through 20.**

Retrograde

Crude Oil is influenced by Mercury retrograde and Venus retrograde.

For 2023, Mercury will be:

- ☼ **retrograde from December 29, 2022 through February 17**

- ☼ **retrograde from April 21 through May 14**

- ☼ **retrograde from August 23 through September 14**

- ☼ **retrograde from December 13 through January 1, 2024.**

For 2023, Venus will be:

- ☼ **retrograde from July 23 to September 3.**

Declination

At the 1983 first trade date, Moon was within 5 degrees of being at its 0-degree declination point. A look at a Crude Oil chart suggests that monthly times of Moon at or near 0 declination are potential times for short term trend changes. This tends to be especially true as Moon is descending from maximum declination down through zero declination. Figure 14-29 illustrates Oil price action in the first part of 2022 with declination events.

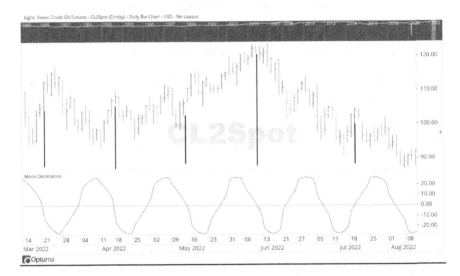

Figure 14-29
Crude Oil and Moon Declination

Cotton

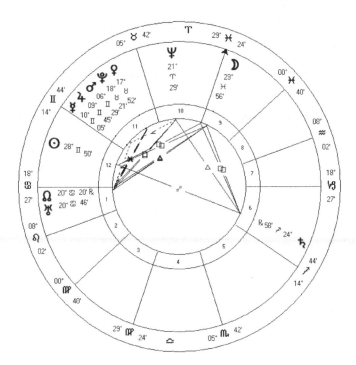

Figure 14-30
Cotton futures First Trade horoscope

After sifting through back-editions of New York newspapers, I have come to conclude that Cotton futures first started trading on June 20, 1870. The horoscope wheel in Figure 14-30 illustrates planetary placements at that time. At first glance, I find it peculiar that the Moon is at the same degree and sign location (24 Pisces) as is the Mid-Heaven in the New York Stock Exchange natal horoscope wheel from 1792. If one assumes that at 6:00 a.m. a trade was initiated, the Mid-Heaven (MH) point is at 24 Pisces. Furthermore, Mars, Jupiter and Mercury are clustered around the location 9 Gemini which is where one finds Uranus in the USA 1776 natal chart. Surely the selection of this Cotton first trade date is no accident.

Outer Planet Aspects

☼ **Saturn will be square the natal Jupiter point in May and early June, 2023.**

☼ **Jupiter will be square the Node in May.**

Sun-Natal Sun Transits

Events of transiting Sun passing 0 and 90-degrees to the natal Sun position are an effective tool for traders to use when navigating the choppy waters of Cotton prices. The Cotton price chart in Figure 14-31 illustrates price action to October 2022. Note the alarming plunge in Cotton prices in June 2022 at a Sun conjunction to natal Sun event.

Figure 14-31
Cotton futures and Sun/natal Sun aspects

In 2023, transiting Sun will aspect natal Sun as follows:

☼ **Transiting Sun will pass 90-degrees to natal Sun from March 7 through March 23**

☼ **Transiting Sun will pass 0-degrees conjunct to natal Sun from June 10 through June 30**

☼ **Transiting Sun will pass 90-degrees to natal Sun from September 13 through September 30**

☼ **Transiting Sun will pass 180-degrees to natal Sun from December 12 through December 28.**

Venus and Natal Moon

One other astro phenomenon traders may wish to consider as a tool to use is the occurrence of Venus passing by the natal Moon position at 24 Pisces.

The price chart in Figure 14-32 illustrates how in May 2022, Cotton prices peaked at a Venus-natal Moon aspect. In July 2022, a square aspect aligned to an important price low.

In 2023, Venus will be conjunct the natal Moon in mid-February.

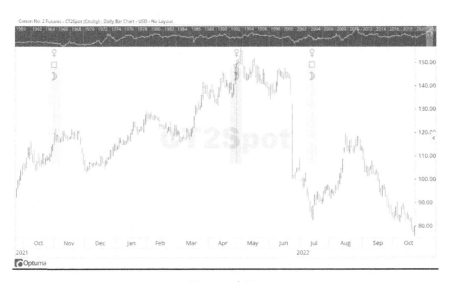

Figure 14-32
Cotton futures and Venus/natal Moon aspects

Declination

At the 1870 first trade date, Mars was very near its declination maximum. Venus was at 15-degrees declination.

A back-study of Cotton prices shows a good correlation between trend changes and Mars declination maxima. In early April 2021, Mars recorded its declination maximum. This event marked the start of a rally that took cotton prices from 79 cents to $1.55 per pound.

In 2023:

- ☼ **the first 3 months of the year will have Mars at maximum declination**
- ☼ **Venus will visit 15-degrees declination in early March and again in late June, 2023**

Coffee

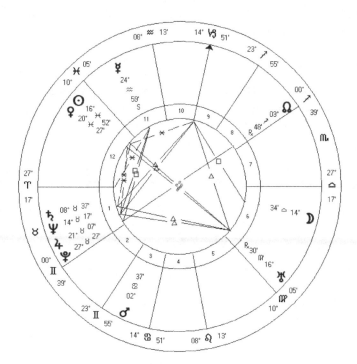

Figure 14-33
Coffee futures First Trade horoscope

Coffee futures started trading in New York in early March of 1882. The horoscope wheel in Figure 14-33 illustrates planetary placements at that time.

Outer Planet Natal Aspects

☼ **From late May through early September 2023, Jupiter will be conjunct to the natal Saturn and Neptune points.**

Sun-Natal Sun Transits

Coffee prices react to transits of Sun passing 0 and 90-degrees to the 1882 natal Sun position at 16 Pisces.

In 2022, transiting Sun passed 0-degrees conjunct to natal Sun from February 29 through March 16. This transit added some energy to the significant price drawdown.

Transiting Sun passed 90-degrees conjunct to natal Sun from May 31 through June 17. This transit created a swing high and the start of what would be a significant drawdown in price. Figure 14-34 illustrates further.

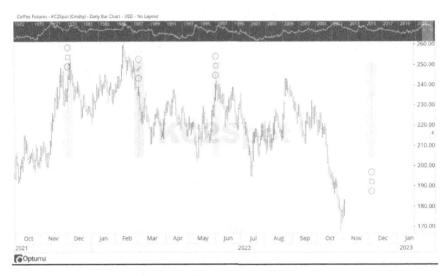

Figure 14-34
Coffee futures and Sun natal Sun transits

Transiting Sun will pass 90-degrees to natal Sun from December 10 through December 16, 2022. At this time of writing, this event is still five weeks away.

For 2023:

☼ **transiting Sun will pass conjunct the 1882 natal Sun location February 29 through March 16, square the natal Sun June 2 through June 13, and square again December 5 through 13.**

Declination

At the 1882 first trade date, Venus, Moon and Sun were all at or near 0 declination. Mars was at its maximum declination.

A look back in time across Coffee prices shows a clear alignment of trend changes to Venus being near 0-degrees declination and Mars being at or near its maximum declination.

Venus was near 0-degrees declination around May 12, 2022. This contributed to a swing low and subsequent rally that took Coffee from $2.03 per pound to $2.40 per pound. Venus again passed 0-degrees declination at October 1, 2022. Coffee prices suffered a significant pullback.

For 2023:

☼ **Mars will be at maximum declination January through March**

☼ **Venus will pass 0 declination around February 23 and November 11.**

Sugar

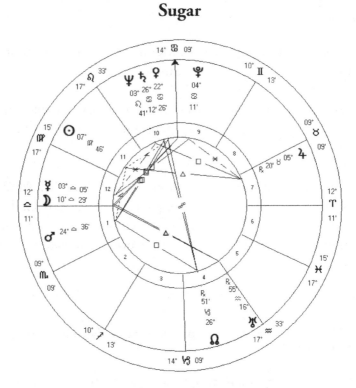

Figure 14-35
Sugar Futures first trade horoscope

Sugar as a bulk commodity started trading in New York as early as 1881. Old editions of New York newspapers suggest that in September 1914 there were plans to open a formal Sugar Exchange, but these plans were scuttled by World War I. Part-way through the War, a formal Exchange *did* open on August 31, 1916. The horoscope wheel in Figure 14-35 illustrates planetary placements at that time. What stands out on this chart wheel is the T-Square formation with Mars at its apex. As well, the Mid Heaven (MH) is at 14 of Cancer is one assumes that the very first trade was conducted at 8:20 am.

184

Outer Planet Natal Aspects

☼ **In June, 2023 Jupiter will be conjunct the natal Jupiter point.**

☼ **From January through late April, Pluto will visit conjunct the natal Node.**

Natal Transits

Events of transiting heliocentric Mars making 0 and 90-degree aspects to the natal heliocentric Mars location (22 Scorpio) have a good propensity to align to pivot swing points. This is the intriguing part of astrology. Even though a significant feature such as a T-square might be noted on a geocentric horoscope, sometimes it is a heliocentric event that triggers action.

For 2022, transiting heliocentric Mars completed a 0-degree aspect in early January. This marked a significant swing low, as Figure 14-36 shows. Heliocentric Mars also made a 90-degree aspect to natal Mars in May which triggered a sizeable price decline.

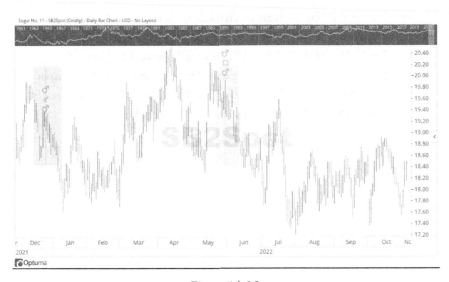

Figure 14-36
Sugar Futures and heliocentric Mars transits

Mercury Retrograde

Mercury retrograde events have a propensity to align to short-term trend changes on Sugar price.

For 2023, Mercury will be:

- ☿ retrograde from December 29, 2022 through February 17

- ☿ retrograde from April 21 through May 14

- ☿ retrograde from August 23 through September 14

- ☿ retrograde from December 13 through January 1, 2024.

Declination

At the 1916 first trade date, Moon was within a whisker of being at 0-degrees of declination. A look back at past price data for Sugar shows that this connection still aligns to price inflection points.

For 2023:

- ☿ **Moon will be at zero declination: January 13, January 26, February 9, February 23, March 8, March 22, April 5, April 18, May 2, May 15, May 29, June 11, June 26, July 8, July 23, August 5, August 19, September 1, September 15, September 29, October 13, October 26, November 9, November 23, December 6, and December 20.**

Cocoa

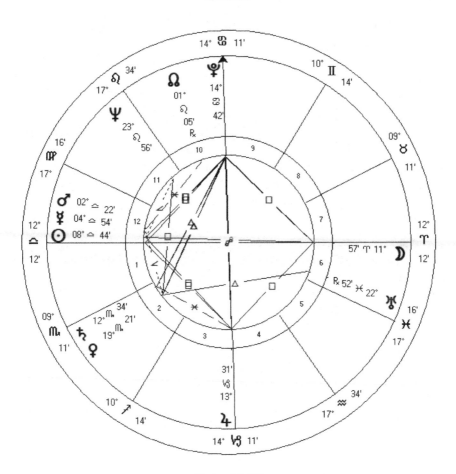

Figure 14-37
Cocoa futures first trade horoscope

Cocoa futures started trading in New York in early October 1925. The horoscope in Figure 14-37 shows planetary placements at the first trade date. What I find peculiar on this horoscope wheel is the Mid-Heaven point located at 14 of Cancer, that same mysterious point that appears in the First Trade horoscope of the New York Stock Exchange.

Outer Planet Natal Aspects

☼ **From January through late March 2023, Jupiter will be square the natal Pluto point.**

Mercury Retrograde

Mercury retrograde events have a high propensity to align to pivot swing points on Cocoa price. The price chart in Figure 14-38 illustrates further. In 2022, the retrograde event aligned to a sharp V-type reversal. The end of retrograde in May saw Cocoa prices take a leg down. The September event saw another V-bottom reversal pattern.

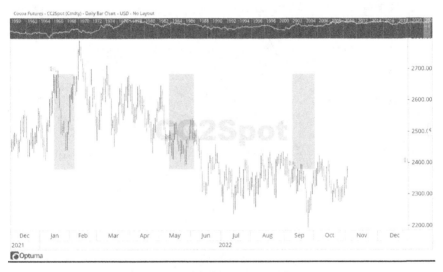

Figure 14-38
Mercury retrograde and Cocoa price

For 2023, Mercury will be:

☼ **retrograde from December 29, 2022 through February 17**

☼ **retrograde from April 21 through May 14**

☼ **retrograde from August 23 through September 14**

☼ **retrograde from December 13 through January 1, 2024.**

Conjunctions and Elongations

The 1925 natal horoscope shows Sun and Mercury conjunct (0-degrees apart). My research has shown that events of Mercury being at its maximum easterly and westerly elongations and events of Mercury being at its Inferior and Superior conjunctions align quite well to pivot swing points.

The price chart in Figure 14-39 illustrates some elongation events that occurred in the first half of 2022. The alignment to pivot price points is intriguing.

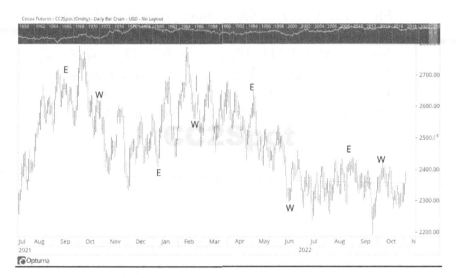

Figure 14-39
Cocoa price and Mercury elongation

☼ **For 2023, Mercury will be at its greatest easterly elongation April 11, August 10, and December 4**

☼ **For 2023, Mercury will be at its greatest westerly elongation January 30, May 29, and September 22.**

Declination

At the 1925 first trade date, Sun, Moon and Mars were all at or near their 0-degree declination levels. A look back at past price data for Cocoa shows that this connection still holds.

For 2023:

☼ Mars will be 0-degrees declination level around August 29.

☼ Sun will be at 0-degrees declination March 20 and September 22.

☼ Moon will be at zero declination: January 13, January 26, February 9, February 23, March 8, March 22, April 5, April 18, May 2, May 15, May 29, June 11, June 26, July 8, July 23, August 5, August 19, September 1, September 15, September 29, October 13, October 26, November 9, November 23, December 6, and December 20.

Feeder Cattle

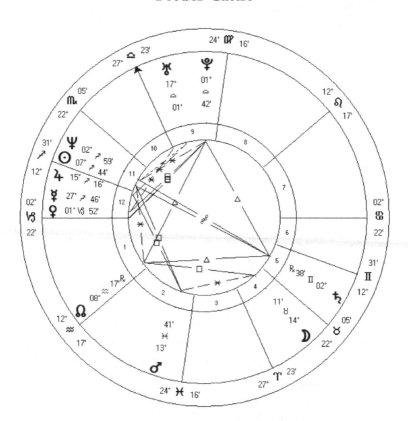

Figure 14-40
Feeder Cattle first trade horoscope

Feeder Cattle futures started trading in Chicago on November 30, 1971. The horoscope in Figure 14-40 shows planetary placements at the first trade date.

Looking carefully at the placements, one can see that Pluto, Saturn, and Node form a triangular pattern. Figure 14-41 illustrates Feeder prices with events of Sun passing the corners of this triangle overlaid on the chart. The Sun-natal Node aspect in January sparked a brief price rally. The Sun-natal Saturn aspect in May caused price to gap higher. The Sun-natal Pluto aspect in September added to the tenor of the price drawdown.

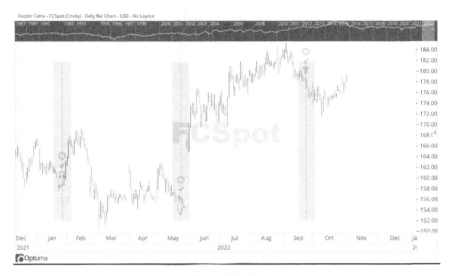

Figure 14-41
Feeder price and Sun aspects

Lastly, (assuming a 9:00 a.m. start to trading) I doubt it was an accident to have the Immum Coeli point (bottom of horoscope wheel) aligned to 24 Pisces, the natal Mid-Heaven point of the NYSE from 1792.

A look back at past charts to the 1980s shows that times of Venus passing 24 Pisces often align to short term trend swings on Feeder prices. For example, in late April 2022, a Venus-24 Pisces conjunction stimulated a rally in Feeder prices which promptly faded once the transit completed.

In 2023;

> ☼ **Venus will pass 24 Pisces in mid-February.**

Declination

At the 1971 first trade date, Venus was very close to its minimum declination. A look back at past chart patterns shows that Venus ast minimum (and also maximum) declination bears a good alignment to points of short-term trend change on Feeder prices. Venus will not record a minimum declination in 2023.

Live Cattle

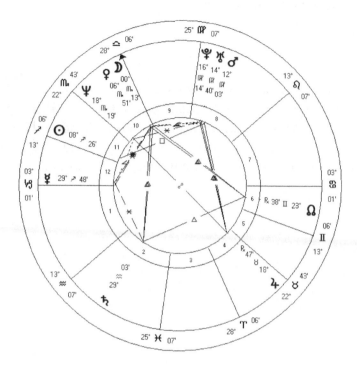

Figure 14-42
Live Cattle first trade horoscope

Live Cattle futures started trading in Chicago on November 30, 1964. The horoscope in Figure 14-42 shows planetary placements at the first trade date.

The most puzzling feature of this horoscope is that it comes at the same day (November 30) as the Feeder Cattle first trade date. Again, note the alignment of the IC point to 24 Pisces. A look back at past charts to the 1960s shows that times of Venus passing 24 Pisces often align to short term trend swings on Live Cattle prices.

Outer Planet Natal Aspects

☼ **From mid-July through early September 2023, Jupiter will be conjunct the natal Jupiter point.**

193

☼ **From February through late April, Saturn will be conjunct the natal Saturn point.**

Declination

At the 1964 first trade date, Venus was at its minimum declination, and Moon was near 0-degrees declination. A look back at past chart patterns shows that Venus minimum (and maximum) declinations bear a good alignment to points of short-term trend change on Live Cattle prices. Unfortunately, in 2023, Venus will not record a minimum declination.

Lean Hogs

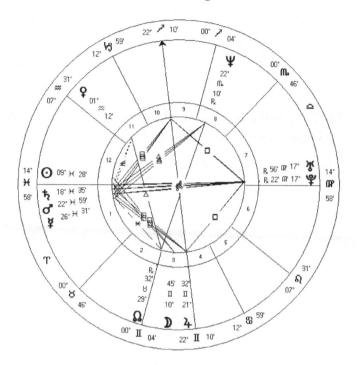

Figure 14-43
Lean Hogs first trade horoscope

Lean Hog futures started trading in Chicago on February 28, 1966. The horoscope in Figure 14-43 shows planetary placements at the first trade date. The most illuminating feature of this horoscope is the appearance of a large square pattern (assuming a 6:45 a.m. trade start) with corners at Saturn, Jupiter, Pluto, and the Mid-Heaven. A look back at past chart data shows a decent alignment to price swing points when Mars and Venus pass the corner points of the square. Be cautioned, however, that Lean Hogs are extremely volatile and not well-suited for risk-averse traders.

Declination

At the 1966 first trade date, Mercury was within 1 degree of being at 0-degrees declination. Mars was within 3 degrees of being at 0-degrees declination. A look back at past chart patterns shows a decent alignment between Mercury 0-degree declination points and price trend swings. A similar observation holds for Mars. But these alignments do not erase the fact that Hogs are volatile to trade.

For 2023:

☼ **Mercury will pass through 0-degrees of declination March 8, August 23-29, and October 8.**

☼ **Mars will pass 0-degrees declination at August 30.**

Lumber

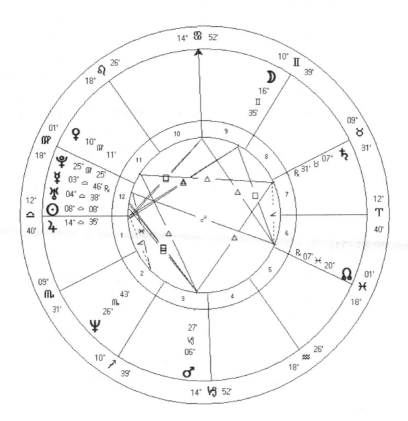

Figure 14-44
Lumber first trade horoscope

Lumber futures started trading in Chicago on October 1, 1969. The horoscope in Figure 14-44 shows planetary placements at the first trade date. Assuming a 7:15 a.m. first trade transaction, the Mid Heaven point is at 14 degrees of Cancer.

Outer Planet Natal Aspects

  ☼ **From late May through early June 2023, Jupiter will be conjunct natal Saturn.**

Mercury Retrograde

The most illuminating feature of this chart is the fact that Mercury was retrograde. If trading lumber futures holds appeal for you, pay close attention to Mercury retrograde events as they may well align to price swing points. For example, in May 2021, Lumber made headline news when prices reached an unprecedented price level of $1400. Within two weeks of this peak, Mercury turned retrograde and with that a swift price drawdown began. Mercury was retrograde from January 13 to February 3, 2022. During this event, Lumber prices fell from $1200 to $826. A few days prior to the start of another retrograde on May 10, 2022, Lumber prices again started to fall.

For 2023, Mercury will be:

- ✿ **retrograde from December 29, 2022 through February 17**
- ✿ **retrograde from April 21 through May 14**
- ✿ **retrograde from August 23 through September 14**
- ✿ **retrograde from December 13 through January 1, 2024.**

Declination

At the 1969 first trade date, Mars was at its declination minimum and Moon was at its declination maximum. Looking back at the historically significant price high in May 2021 shows that Mars was at its declination maximum (the opposite extreme from the 1969 first trade date), while Moon was at 0 declination.

For 2023:

- ✿ **Mars will be at maximum declination for January through March.**

Platinum

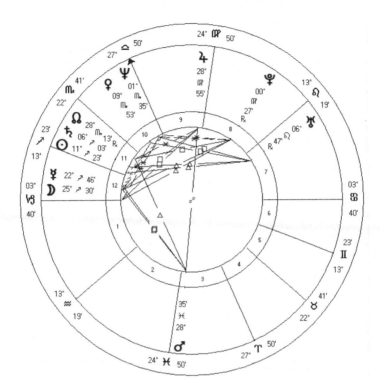

Figure 14-45
Platinum first trade horoscope

Platinum futures started trading in New York on December 3, 1956. Assuming an 8:50 a.m. first trade transaction, note the alignment of the IC point to 24 Pisces, which happens to also be the NYSE natal Mid-Heaven point.

Declination

At the 1956 first trade date, Moon was at its minimum declination and Mercury was not only at a declination low, it was what astrologers call *out-of-bounds (OB)*. A planet is OB when its declination maximum or minimum exceeds by a few degrees what its maxima or minima normally would be.

Mercury was at a declination minimum in mid-December 2021. This stimulated a Platinum rally from $910 to $1175 per ounce.

For 2023:

☼ **Mercury will exhibit minimum declination in early December.**

Palladium

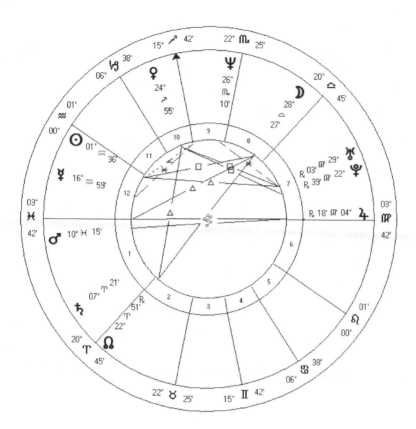

Figure 14-46
Palladium first trade horoscope

Palladium futures started trading in New York on January 22, 1968.

Outer Plant Natal Aspects

☼ **In January and February, Jupiter will pass conjunct to the natal Saturn point.**

Declination

At that date, Moon was within a degree of being at 0-degrees declination. Venus was at its minimum declination. A look back at past chart patterns shows that Venus declination minima and Moon at 0-degrees declination do align to the various rallies that have unfolded over the years.

For 2023, Venus will not record an actual declination low.

Natural Gas

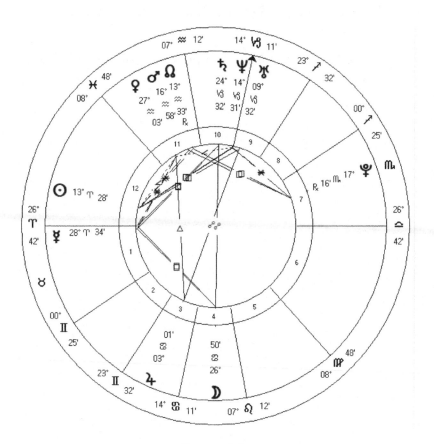

Figure 14-47
Natural Gas first trade horoscope

Natural Gas futures started trading in Chicago on April 3, 1990. At that date, Moon had just made its declination maximum. No other planetary declination level stands out as being curiously interesting. The above horoscope does however have that always intriguing 14 of Cancer point at the horoscope IC point. A look back at past chart data suggests that times when Mars, Venus, and Sun variously transit past 14 Cancer, often align to price swing points. In early July 2022, Natural Gas had suffered a sizeable drawdown in price. A V-bottom formation came just as Sun was passing 14 Cancer.

Outer Planet Natal Aspects

- ☼ In the first part of January, Jupiter will pass square to the natal Jupiter point.

- ☼ In the first part of March, Jupiter will pass the natal Sun point.

CHAPTER FIFTEEN
Quantum Science

Quantum Price Lines

Quantum Price Lines are based on Einstein's quantum theory. The notion of Quantum Lines posits that the price of a stock, index or commodity can be thought of as a light particle or electron that can occupy different energy levels or orbital shells.

Author and market researcher Fabio Oreste combined the notion of Quantum Price Lines with Einstein's theory that the fabric of space-time can be bent. Picture a group of people holding the edges of a large blanket. They pull on the edges until the blanket is stretched tight. Next, someone places a ball on the tight blanket. The weight of the ball causes a slight sag in the blanket fabric. Oreste says the point of maximum curvature of the sagged portion is akin to a Quantum Price Line. In his book entitled *Quantum Trading*, (1) Oreste details his formula for Quantum Price Line calculation:

Quantum Line = (N x 360) + PSO ;

Where N is the harmonic level = 1,2,3,4,5,6,8,...

Always think of harmonics in terms of divisions of a circle. A 3rd harmonic (N=3) is 120-degrees. A 5th harmonic is 360/5 =72 degrees. And so on.

Where PSO = heliocentric planetary longitude x Conversion Scale

The Conversion Scale = 2^n ; 1,2,4,8,16,...; where n=0,1,2,3,4,....

When dealing with prices less than 360, the inverse variation of the formula is used.

Quantum Line = (1/N x 360) + PSO

Oreste's technique then allows one to calculate various sub-divisions of these Quantum Lines. Taking the value of the calculated Quantum Line, one would generate the sub-divisions by multiplying by 1.0625, 1.125, 1.875, 1.25, etc... in steps of 0.0625.

Please note the use of *heliocentric* planetary data in these Quantum Line calculations. There are websites that will provide you with this data such as **www.astro.com/swisseph**. Alternatively, you can purchase a Heliocentric Ephemeris book such as *The American Heliocentric Ephemeris, 2001-2050*.

To assist you with calculating Quantum Lines, consider the following example:

On a given date, suppose the following heliocentric planetary positions are noted: Mars 306 degrees, Jupiter 307 degrees, Neptune 324 degrees, Pluto 271 degrees.

In this example, let N=1 and let the conversion scale be set to CS=1. PSO will be the planetary longitude x CS.

The Oreste point of maximum curvature for these planets is then:

Mars: (N x 360) + PSO; which is (1 x360) + 306 = 666
Jupiter: 360 + 307 = 667
Neptune: 360 + 324 = 684
Pluto: 360 + 271 = 631

If you were to take another date in the future and calculate the points of maximum curvature, you could then join the two points for each planet. By definition two points joined equals a straight line. You could then extend these lines out into the future. These lines are called *Quantum Lines* (or QL's).

If the above maximum curvature numbers seem oddly familiar, think to early 2009 and consider: The S&P 500 at the March 6th, 2009 lows delivered an intra-day low of 665.7 and on the day the close was 687. Indeed, Mars, Jupiter and Neptune all acted in concert in March 6, 2009 to provide a floor of support under the US equity market.

For larger price values such as those in an equity index, it becomes necessary to use a bigger value for N. Consider the following example of the S&P 500. The high on the S&P 500 at September 7, 2021 was 4546. Intuitively, a larger number such as this will demand something more than N=1 and CS=1. The question is, are there Oreste points of maximum curvature at or near this level that would suggest the S&P 500 had hit a significant resistance level?

The first step in answering this question is to obtain the heliocentric planetary data for the date in question. Positions of planets were:

Mars at 180-degrees, Jupiter at 328, Neptune at 351, Saturn at 311, Pluto at 295.

Next, choose a CS value. Given the magnitude of the S&P 500, I am going to pick CS=2^3 = 8.

The Quantum Line formula is:

(N x 360) + (heliocentric position x conversion scale)

Calculating the back half of the formula gives us:

Mars: 8x180 = 1440
Jupiter: 8x328 = 2624
Neptune: 8x351 = 2808
Saturn: 8x311 = 2488
Pluto: 8x295 = 2360

Next, a harmonic N value is required. Consider the 5th harmonic of N=5.

The formula then yields:

Mars: (5x360) + 1440 = 3240
Jupiter: (5x360) + 2624 = 4424
Neptune: (5x360) + 2808 = 4608
Saturn: (5x360) +2488 = 4288
Pluto: (5x360) + 2360 = 4160

From these numbers we can conclude that a Neptune 5th harmonic at 4608 was foretelling of overhead resistance on September 7, 2021.

Next, consider the 6th harmonic and N=6x360 = 2160.

The recalculated values become:

Mars: (6x360) + 1440 = 3600
Jupiter: (6x360) + 2624 = 4784
Neptune: (6x360) + 2808 = 4968
Saturn: (6x360) + 2488 = 4648
Pluto: (6x360) + 2360 = 4520

From these numbers we can conclude that Pluto 6th harmonic at 4520 was acting as resistance on September 7th. A Saturn 6th harmonic was situated close by.

What follows is a suggested list of some points of maximum curvature you can apply to various indices and commodities for 2023.

These various points are based on the following heliocentric planetary positions at January 1, July 1, and December 30, 2023. As 2023 starts, sketch the Jan 1 and July 1 points on your charts. Join the points with a line. Later in 2023, add the December 30 point and extend your lines.

HELIOCENTRIC DEGREE POSITION			
Planet	Jan 1, 2023	July 1, 2023	Dec 30, 2023
Jupiter	12	29	43
Saturn	326	332	335
Neptune	354	355	356
Uranus	47	49	50
Pluto	298	299	299

2023 Quantum Levels

S&P 500 Index (CS=128)

PLANET & HARMONIC	JAN 1	JULY 1	DEC 30
Jupiter 1st	1896	4072	5864
Jupiter 2nd	2256	4432	6224
Jupiter 3rd	2616	4792	-
Jupiter 4th	2976	5152	-
Jupiter 5th	3336	5512	-
Jupiter 6th	3896	5872	-
Jupiter 8th	4416	-	-
Jupiter 9th	4776	-	-
Jupiter 10th	5136	-	-
Jupiter 12th	5856	-	-

S&P 500 Index (CS=64)

PLANET & HARMONIC	JAN 1	JULY 1	DEC 30
Jupiter 1st	-	-	3112
Jupiter 2nd	-	-	3472
Jupiter 3rd	-	2936	3832
Jupiter 4th	-	3296	4192
Jupiter 5th	-	3656	4552
Jupiter 6th	2928	4016	4912
Jupiter 8th	3648	4736	5632
Jupiter 9th	4008	5096	-
Jupiter 10th	4368	5456	-
Jupiter 12th	5088	-	-

S&P 500 Index (CS=64)

PLANET & HARMONIC	JAN 1	JULY 1	DEC 30
Uranus 1st	3368	3496	3560
Uranus 2nd	3728	3856	3920
Uranus 3rd	4088	4216	4280
Uranus 4th	4448	4576	4640
Uranus 5th	4808	4936	5000
Uranus 6th	5168	5296	5360
Uranus 8th	5888	6016	6080

S&P 500 Index (CS=8)

PLANET & HARMONIC	JAN 1	JULY 1	DEC 30
Saturn 2nd	3328	3376	3400
Saturn 3rd	3688	3736	3760
Saturn 4th	4048	4096	4120
Saturn 5th	4408	4456	4480
Saturn 6th	4768	4816	4840
Saturn 8th	5488	5536	5560
Neptune 2nd	3552	3560	3568
Neptune 3rd	3912	3920	3928
Neptune 4th	4272	4280	4288
Neptune 5th	4632	4640	4648
Neptune 6th	4996	5000	5008
Neptune 8th	5716	5720	5728
Pluto 2nd	3104	3112	3112
Pluto 3rd	3464	3472	3472
Pluto 4th	3824	3832	3832
Pluto 5th	4184	4192	4192
Pluto 6th	4544	4552	4552
Pluto 8th	5264	5272	5272

NASDAQ Composite Index (CS=16)

PLANET & HARMONIC	JAN 1	JULY 1	DEC 30
Saturn 12th	9536	9632	9680
Saturn 15th	10616	10712	10760
Saturn 16th	10976	11072	11120
Neptune 12th	9984	10000	10016
Neptune 15th	11064	11080	11096
Neptune 16th	11424	11440	11456
Pluto 12th	9088	9104	9104
Pluto 15th	10168	10184	10184
Pluto 16th	10528	10544	10544

NASDAQ Composite Index (CS=32)

PLANET & HARMONIC	JAN 1	JULY 1	DEC 30
Saturn 2nd	11152	11344	12112
Saturn 3rd	11512	11704	12472
Saturn 4th	11872	12064	12832
Saturn 5th	12232	12424	13192
Saturn 6th	12592	12784	13552
Saturn 8th	13312	13504	14272
Saturn 9th	13672	13864	14632
Saturn 10th	14032	14224	14992
Saturn 12th	14752	14944	15712
Saturn 15th	15832	16024	16792
Saturn 16th	16192	16384	17152
Saturn 18th	16912	17104	17872

NASDAQ Composite Index (CS=32)

PLANET & HARMONIC	JAN 1	JULY 1	DEC 30
Neptune 1st	11688	11720	11752
Neptune 2nd	12048	12080	12112
Neptune 3rd	12408	12440	12472
Neptune 4th	12768	12800	12832
Neptune 5th	13128	13160	13192
Neptune 6th	13488	13520	13552
Neptune 8th	14208	14240	14272
Neptune 10th	14928	14960	14992
Neptune 12th	15648	15680	15712
Neptune 15th	16728	16760	16792
Neptune 16th	17088	17120	17152
Neptune 18th	17808	17840	17872

NASDAQ Composite Index (CS=256)

PLANET & HARMONIC	JAN 1	JULY 1	DEC 30
Jupiter 1st	-	-	11368
Jupiter 2nd	-	-	11728
Jupiter 3rd	-	-	12088
Jupiter 4th	-	-	12448
Jupiter 5th	-	-	12808
Jupiter 6th	-	-	13168
Jupiter 8th	-	11344	13888
Jupiter 9th	-	11704	14248
Jupiter 10th	-	12064	14608
Jupiter 12th	-	12784	15328
Jupiter 15th	-	13864	16408
Jupiter 16th	-	14224	16768
Jupiter 18th	-	14944	17488
Jupiter 20th	10272	14624	18208
Jupiter 24th	11712	16064	19648
Jupiter 30th	13872	18224	-
Jupiter 36th	16032	-	-

NASDAQ Composite Index (CS=32)

PLANET & HARMONIC	JAN 1	JULY 1	DEC 30
Pluto 1st	9896	9928	9928
Pluto 2nd	10256	10288	10288
Pluto 3rd	10616	10648	10648
Pluto 4th	10976	11008	11008
Pluto 5th	11336	11368	11368
Pluto 6th	11696	11728	11728
Pluto 8th	12416	12448	12448
Pluto 9th	12776	12808	12808
Pluto 10th	13136	13168	13168
Pluto 12th	13856	13888	13888
Pluto 15th	14936	14968	14968
Pluto 16th	15296	15328	15328

FTSE 100 Index (CS=128)

PLANET & HARMONIC	JAN 1	JULY 1	DEC 30
Jupiter 12th	5856	8032	-
Jupiter 6th	-	5872	7664
Jupiter 8th	4416	6592	8384

FTSE 100 Index (CS=16)

PLANET & HARMONIC	JAN 1	JULY 1	DEC 30
Saturn 1st	5856	8032	-
Saturn 2nd	5936	6032	6080
Saturn 3rd	6296	6392	6440
Saturn 4th	6656	6752	6800
Saturn 5th	7016	7112	7160
Saturn 6th	7376	7472	7520
Saturn 8th	8096	8192	8240

FTSE 100 Index (CS=16)

PLANET & HARMONIC	JAN 1	JULY 1	DEC 30
Neptune 1st	6024	6040	6056
Neptune 2nd	6384	6400	6416
Neptune 3rd	6744	6760	6776
Neptune 4th	7104	7120	7136
Neptune 5th	7464	7480	7496
Neptune 6th	7824	7840	7856
Neptune 8th	8455	8560	8576

FTSE 100 Index (CS=16)

PLANET & HARMONIC	JAN 1	JULY 1	DEC 30
Pluto 3rd	5848	5864	5864
Pluto 4th	6208	6224	6224
Pluto 5th	6568	6584	6584
Pluto 6th	6928	6944	6944
Pluto 8th	7648	7664	7664
Pluto 9th	8008	8024	8024

S&P ASX 200 Index (CS=16)

PLANET & HARMONIC	JAN 1	JULY 1	DEC 30
Saturn 1st	5856	8032	-
Saturn 2nd	5936	6032	6080
Saturn 3rd	6296	6392	6440
Saturn 4th	6656	6752	6800
Saturn 5th	7016	7112	7160
Saturn 6th	7376	7472	7520
Saturn 8th	8096	8192	8240

S&P ASX 200 Index (CS=16)

PLANET & HARMONIC	JAN 1	JULY 1	DEC 30
Neptune 1st	6024	6040	6056
Neptune 2nd	6384	6400	6416
Neptune 3rd	6744	6760	6776
Neptune 4th	7104	7120	7136
Neptune 5th	7464	7480	7496
Neptune 6th	7824	7840	7856
Neptune 8th	8455	8560	8576

S&P ASX 200 Index (CS=16)

PLANET & HARMONIC	JAN 1	JULY 1	DEC 30
Pluto 3rd	5848	5864	5864
Pluto 4th	6208	6224	6224
Pluto 5th	6568	6584	6584
Pluto 6th	6928	6944	6944
Pluto 8th	7648	7664	7664
Pluto 9th	8008	8024	8024

Gold Futures (CS=2)

PLANET & HARMONIC	JAN 1	JULY 1	DEC 30
Saturn 3rd	1732	1744	1750
Saturn 4th	2092	2104	2110
Saturn 5th	2452	2464	2470
Neptune 3rd	1788	1790	1792
Neptune 4th	2148	2150	2152
Neptune 5th	2508	2510	2512
Pluto 3rd	1676	1678	1678
Pluto 4th	2036	2038	2038
Pluto 4th	2396	2398	2398

Gold Futures (CS=2)

PLANET & HARMONIC	JAN 1	JULY 1	DEC 30
Jupiter 3rd	1464	2008	2456
Jupiter 4th	1824	2368	2816
Jupiter 5th	2184	2728	3176
Jupiter 6th	2544	3536	-

Silver Futures (CS=1/64)

PLANET & HARMONIC	JAN 1	JULY 1	DEC 30
Jupiter 12th	$30.66	$31.60	$32.38
Jupiter 16th	$23.17	$24.11	$24.89
Jupiter 18th	$20.68	$21.63	$22.41
Jupiter 20th	$18.67	$19.61	$20.39
Saturn 45th	$26.12	$26.45	$26.22
Saturn 60th	$24.14	$24.47	$24.64
Neptune 36th	$29.69	$29.75	$29.80
Neptune 40th	$28.68	$28.74	$28.79
Neptune 45th	$27.67	$27.73	$27.79
Pluto 18th	$34.57	$34.62	$34.62
Pluto 20th	$22.65	$22.66	$22.66
Pluto 24th	$19.66	$19.68	$19.68
Pluto 30th	$16.64	$16.65	$16.65

Currency Futures: Canadian Dollar, Australian Dollar (CS=1/1024)

PLANET & HARMONIC	JAN 1	JULY 1	DEC 30
Saturn 5th	$1.03	$1.04	$1.05
Saturn 6th	$0.91	$0.91	$0.92
Saturn 8th	$0.768	$0.774	$0.777
Saturn 10th	$0.678	$0.684	$0.687
Neptune 5th	$1.06	$1.06	$1.07
Neptune 6th	$0.94	$0.94	$0.94
Neptune 8th	$0.795	$0.795	$0.795
Neptune 10th	$.705	$0.71	$0.71
Pluto 4th	$1.19	$1.19	$1.19
Pluto 5th	$1.00	$1.01	$1.01
Pluto 6th	$0.89	$0.89	$0.89
Pluto 8th	$0.741	$0.741	$0.741
Pluto 10th	$0.651	$0.651	$0.651

Currency Futures: Euro and British Pound (CS=1/512)

PLANET & HARMONIC	JAN 1	JULY 1	DEC 30
Jupiter 3rd	$1.20	$1.22	$1.24
Saturn 6th	$1.33	$1.35	$1.35
Saturn 8th	$1.21	$1.23	$1.23
Saturn 9th	$1.06	$1.08	$1.08
Neptune 6th	$1.41	$1.41	$1.41
Neptune 8th	$1.29	$1.29	$1.29
Neptune 9th	$1.14	$1.14	$1.14
Pluto 6th	$1.30	$1.30	$1.30
Pluto 8th	$1.18	$1.18	$1.18
Pluto 9th	$1.03	$1.03	$1.03

Wheat and Corn Futures (CS=1/64)

PLANET & HARMONIC	JAN 1	JULY 1	DEC 30
Saturn 1st	$8.69	$8.78	$8.83
Saturn 2nd	$6.89	$6.98	$7.03
Saturn 4th	$5.99	$6.08	$6.13
Neptune 1st	$9.13	$9.14	$9.16
Neptune 2nd	$7.33	$7.34	$7.36
Neptune 4th	$6.43	$6.44	$6.46
Pluto 1st	$8.25	$8.27	$8.27
Pluto 2nd	$6.45	$6.47	$6.47

Wheat and Corn Futures (CS=1/32)

PLANET & HARMONIC	JAN 1	JULY 1	DEC 30
Saturn 2nd	$11.98	$12.17	$12.26
Saturn 3rd	$11.38	$11.57	$11.66
Saturn 4th	$10.98	$11.27	$11.36
Saturn 5th	$10.90	$11.09	$11.18
Saturn 6th	$10.78	$10.97	$11.06
Saturn 8th	$10.63	$10.82	$10.91
Saturn 10th	$10.54	$10.73	$10.82
Neptune 2nd	$12.86	$12.89	$12.92
Neptune 3rd	$12.26	$12.29	$12.32
Neptune 4th	$11.96	$11.99	$12.02
Neptune 5th	$11.78	$11.81	$11.84
Neptune 6th	$11.66	$11.69	$11.72
Neptune 8th	$11.51	$11.54	$11.57
Neptune 9th	$11.46	$11.49	$11.52
Neptune 10th	$11.42	$11.45	$11.48

Wheat and Corn Futures (CS=1/32)

PLANET & HARMONIC	JAN 1	JULY 1	DEC 30
Pluto 2nd	$11.11	$11.12	$11.12
Pluto 3rd	$10.51	$10.52	$10.52
Pluto 4th	$10.21	$10.22	$10.22
Pluto 5th	$10.03	$10.04	$10.04
Pluto 6th	$9.91	$9.92	$9.92
Pluto 8th	$9.76	$9.77	$9.77
Pluto 10th	$9.67	$9.68	$9.68
Pluto 12th	$9.61	$9.62	$9.62
Pluto 16th	$9.53	$9.54	$9.54
Pluto 18th	$9.51	$9.52	$9.52

Soybean Futures (CS=3)

PLANET & HARMONIC	JAN 1	JULY 1	DEC 30
Saturn 1st	$13.38	$13.56	$13.65
Saturn 2nd	$11.58	$11.76	$11.85
Saturn 3rd	$10.68	$10.86	$10.95
Neptune 1st	$14.22	$14.25	$14.28
Neptune 2nd	$12.42	$12.45	$12.48
Neptune 3rd	$11.82	$11.85	$11.88
Neptune 4th	$11.52	$11.55	$11.58
Pluto 1st	$12.54	$12.57	$12.57
Pluto 2nd	$10.74	$10.77	$10.77

Soybean Futures (CS=4)

PLANET & HARMONIC	JAN 1	JULY 1	DEC 30
Saturn 1st	$16.64	$16.88	$17.00
Saturn 2nd	$14.84	$15.08	$15.20
Saturn 3rd	$14.24	$14.48	$14.60
Saturn 4th	$13.94	$14.18	$14.30
Saturn 5th	$13.76	$14.00	$14.12
Saturn 6th	$13.64	$13.88	$14.00
Saturn 8th	$13.49	$13.73	$13.85
Saturn 10th	$13.40	$13.64	$13.76
Neptune 1st	$17.76	$17.80	$17.84
Neptune 2nd	$15.96	$16.00	$16.04
Neptune 3rd	$15.36	$15.40	$15.44
Neptune 4th	$15.06	$15.10	$15.14
Neptune 5th	$14.88	$14.92	$14.96
Neptune 6th	$14.76	$14.80	$14.84
Pluto 1st	$15.52	$15.56	$15.56
Pluto 2nd	$13.72	$13.76	$13.76
Pluto 3rd	$13.12	$13.16	$13.16
Pluto 4th	$12.82	$12.86	$12.86

Soybean Futures (CS=4)

PLANET & HARMONIC	JAN 1	JULY 1	DEC 30
Saturn 1st	$85.10	$86.60	$87.35
Saturn 2nd	$83.30	$86.60	$87.35
Saturn 3rd	$82.70	$84.20	$84.95
Saturn 4th	$82.40	$83.90	$84.65
Neptune 1st	$92.10	$92.35	$92.60
Neptune 2nd	$90.30	$90.55	$90.88
Neptune 3rd	$89.70	$89.95	$90.20
Neptune 4th	$89.40	$89.65	$89.90
Pluto 1st	$78.10	$78.35	$78.35
Pluto 2nd	$76.30	$76.55	$76.55

Crude Oil Futures (CS=1/2)

PLANET & HARMONIC	JAN 1	JULY 1	DEC 30
Saturn 1st	$166.60	$169.60	$171.10
Saturn 2nd	$164.80	$169.60	$171.10
Neptune 1st	$180.60	$181.10	$181.60
Neptune 2nd	$178.80	$179.30	$179.80
Pluto 1st	$152.60	$153.10	$153.10
Pluto 2nd	$150.80	$151.30	$151.30

CHAPTER SIXTEEN
Conclusion

I have taken you on a wide-ranging journey in this Almanac to acquaint you with the mathematical links between planetary activity and market price behavior. I sincerely hope you will embrace planetary cycles as a valuable tool to assist you in your trading and investing activity. I hope you will pause often to contemplate whether the correlations you have learned about in this Almanac are the actions of the cosmos on the emotions of traders and investors or the actions of power players using the planets to manipulate the markets.

If you decide to embrace financial astrology as a tool to help you navigate the markets, I encourage you to stick with it. At first it might seem daunting, but fight the urge to give up. Soon enough, your trading and investing activity will take on a new meaning.

To encourage you, I will leave you with the words of Neil Turok from his 2012 book, *The Universe Within.* [1]

"Perseverance leads to enlightenment. And the truth is more beautiful than your wildest dreams."

NOTES
&
RECOMMENDED
READING

Notes

Introduction

1)   McWhirter, L. (1938) *McWhirter Theory of Stock Market Forecasting.* Astro Book Company, USA.

2)   Bradley, D. (1948) *Stock Market Prediction.* Llewellyn Publishers, USA.

Chapter 1

Figure 1-1: taken from Loes Ten Kate, I. (2006) ORGANICS ON MARS -Laboratory studies of organic material under simulated Martian conditions.

Figure 1-2: taken from www.wikimediacommons.com. File Earths orbit and ecliptic.PNG

Figure 1-3: taken from https://www.elsaelsa.com/astrology/zodiac-sign-glyphs

Figure 1-4: taken from http://mysticaltransformations.com

Figure 1-5: taken from https://serc.carleton.edu/mel/teaching_resources/moon_mel.html

Figure 1-8: taken from http://www.astronomy.ohio-state.edu/~pogge/Ast161/Unit2

Chapter 3

1) McWhirter, L. (1938) *McWhirter Theory of Stock Market Forecasting*. Astro Book Company, USA.

Chapter 4

Figure 4-1: taken from Falconer, K. (2003) Fractal Geometry: Mathematical Foundations and Applications. John Wiley & Sons, USA.

1) Braden, G. (2009) *Fractal Time*. Hay House, USA.

2) Haulman, C. (2010) The Panic of 1819. [online] Available at: https://www.moaf.org/exhibits/checks_balances/andrew-jackson/materials/Panic_of_1819.pdf. Accessed: November 2021.

3) Richardson, G., Sablik, T. (2015) *Banking Panics of the Guilded Age*. [online] Available at: https://www.federalreservehistory.org/essays/banking-panics-of-the-gilded-age. Accessed: November 2021.

4) Reinhart, C., Rogoff, K. (2009). *This Time is Different: Eight Centuries of Financial Folly*. Princeton University Press.

5) Moen, J., Tallman, E. (2015) *The Panic of 1907*. [online] Available at: https://www.federalreservehistory.org/essays/panic-of-1907. Accessed: November 2021.

6) NASA Eclipse website (2021) *Eclipses and the Saros*. [online] Available at: https://eclipse.gsfc.nasa.gov/SEsaros/SEsaros.html. Accessed: November 2021.

7) NBER website (2021) *US Business Cycle Expansions and Contraction*. [online] Available at: https://www.nber.org/research/data/us-business-cycle-expansions-and-contractions. Accessed: November 2021.

8) Vernon, J. R. (1991). The 1920-21 Deflation: The Role of Aggregate Supply. *Economic Inquiry*. 29 (3),pp: 572–580.

9) Pugesek, B. (2014) Fractal Cycle Turning Points: A Theory of Human Social Progression. *Ecological Complexity*. Vol. 20. pp: 157-75.

Chapter 7

1) Takahashi, F., Shimizu, H., Tsunakawa, H. (2019) Mercury's anomalous magnetic field caused by a symmetry-breaking self-regulating dynamo. *Nature Communications*. 10 (208).

Figure 7-1: taken from http://www.astronomy.ohio-state.edu/~pogge/Ast161/Unit2.

Chapter 10

1) Long, J. (1992) *Basic astrotech: A new technique for trading commodities using geocosmic energy fields with technical analysis*. 6th ed. Professional Astrology Service Inc. USA.

2) Bradley, D. (2004) *Stock Market Prediction: The Historical and Future Siderograph Charts*. Books Work. USA.

Chapter 11

1) Cahn, J. (2011) *The Harbinger*. Charisma Media, USA.

2) Cahn, J. (2016) *The Book of Mysteries*. Charisma Media, USA.

3) Cahn, J. (2017) *The Paradigm*. Charisma Media, USA.

4) Wong, M. (2005) Tunnel Thru the Air, *Trader's World*, Issue 39, p.46

Chapter 12

1) Kramer, J. (1995) *Astrology Really Works*. Hay House, USA.

2) Gann, W.D. (1927) *Tunnel Through the Air*. Pantainos Classics, USA.

Chapter 13

1) McWhirter, L. (1938) *McWhirter Theory of Stock Market Forecasting*. Astro Book Company, USA.

Chapter 15

1) Oreste, F. (2011) *Quantum Trading*. J. Wiley & Sons, USA.

Chapter 16

1) Turok, N. (2012) *The Universe Within*. House of Anansi Press, Canada.

Recommended Readings

The Bull, the Bear and the Planets, M.G. Bucholtz, (USA, 2013)

The Lost Science, M.G. Bucholtz, (USA, 2013)

Stock Market Forecasting – The McWhirter Method De-Mystified, M.G. Bucholtz, (Canada, 2014)

The Cosmic Clock, M.G. Bucholtz, (Canada, 2016)

The Universal Clock, J. Long, (USA, 1995)

A Theory of Continuous Planet Interaction, *NCGR Research Journal*, T. Waterfall, Volume 4, Spring 2014, pp 67-87.

Financial Astrology, Giacomo Albano, (U.K., 2011)

GLOSSARY

Ascendant: one of four cardinal points on a horoscope, the Ascendant is situated in the East

Aspect: the angular relationship between two planets measured in degrees

Autumnal Equinox (see Equinox): – that time of year when Sun is at 0-degrees Libra

Conjunct: an angular relationship of 0-degrees between two planets

Cosmo-biology: changes in human emotion caused by changes in cosmic energy

Declination: the amount (in degrees) that a planet wanders above or below the ecliptic plane

Descendant: one of four cardinal points on a horoscope, the Descendant is situated in the West

Ecliptic Plane: The plane of motion traveled by the planets as they orbit the Sun

Elongation: refers to the angle subtended between a planet and the Sun based on an observers position on Earth

Ephemeris: a daily tabular compilation of planetary and lunar positions

Equinox: an event occurring twice annually, an equinox event marks the time when the tilt of the Earth's axis is neither toward or away from the Sun

Fibonacci Sequence: a recursive mathematical sequence in which a given term is the sum of the two preceding terms. (The infinite sequence is as follows: 0,1,1,2,3,5,8,13,21,34,55,89…)

First Trade chart: a zodiac chart depicting the positions of the planets at the time a company's stock or a commodity future commenced trading on a recognized financial exchange

First Trade date: the date a stock or commodity futures contract first began trading on a recognized exchange

Fractal: a repetitive mathematical pattern first mathematically recognized by mathematician Benoit Mandelbrot

Full Moon: from a vantage point situated on Earth, when the Moon is seen to be 180-degrees to the Sun

Gann Master Cycle: the 19.86 year time span from heliocentric Saturn and Jupiter being conjunct to once again being conjunct

Geocentric: planetary location system in which the vantage point for determining planetary aspects is the Earth

Heliocentric: planetary location system in which the vantage point for determining planetary aspects is the Sun

Horoscope: an image of the zodiac overlaid with the positions of the planets

House: a $1/12^{th}$ portion of the zodiac. Portions are not necessarily equal depending on the mathematical formula used to calculate the divisions

Lucas Sequence: a recursive mathematical sequence in which a given term is the sum of the two preceding terms. (The infinite sequence is as follows: 2,1,3,4,7,11,18,29,47,76…)

Lunar Month: (see Synodic Month)

Lunation: (see New Moon)

McWhirter Cycle: The 18.6 year time span in which the North Node progresses around the 12 zodiac signs

Mid-Heaven: one of four cardinal points on a horoscope, situated in the South

Natal: the position of a planet at a point in history, usually the first trade date

New Moon: from a vantage point situated on Earth, when the Moon is seen to be 0-degrees to the Sun.

North Node of Moon: the intersection points between the Moon's plane and Earth's ecliptic are termed the North and South nodes (Astrologers tend to focus on the North node and Ephemeris tables clearly list the zodiacal position of the North Node for each calendar day.)

Orb: the amount of flexibility or tolerance given to an aspect

Quantum Point: a mathematical construct that refers to the point of maximum distortion of the time-space fabric due to the presence of a planet

Retrograde motion: the apparent backwards motion of a planet through the zodiac signs when viewed from a vantage point on Earth

Sidereal Month: the Moon orbits Earth with a slightly elliptical pattern in approximately 27.3 days, relative to a fixed frame of reference

Sidereal Orbital Period: the time required for a planet to make one full orbit of the Sun as viewed from a fixed vantage point on the Sun.

Siderograph: a mathematical equation developed by astrologer Donald Bradley in 1946 (By plotting the output of the equation against date, inflection points can be seen on the plotted curve. It is at these inflection points that human emotion is most apt to change resulting in a trend change on the Dow Jones or S&P 500 Index)

Solstice: occurring twice annually, a solstice event marks the time when the Sun reaches its highest or lowest altitude above the horizon at noon

Synodic Month: from a moving frame of reference, the 29.5 day time span for the Moon to orbit the Earth

Synodic Orbital Period: the time required for a planet to make one full orbit of the Sun as viewed from a fixed vantage point on Earth

Transiting: the motion of a planet at a given time as identified by data in an Ephemeris table

Vernal Equinox: that time of the year when Sun is at 0-degrees Aries

Zodiac: an imaginary band encircling the 360-degrees of the planetary system divided into twelve equal portions of 30-degrees each

Zodiac Wheel: a circular image broken into 12 portions of 30-degrees each. Each portion represents a different astrological sign

ABOUT THE AUTHOR

Malcolm Bucholtz, B.Sc, MBA, M.Sc., is a graduate of Queen's University (Faculty of Engineering) in Canada and Heriot Watt University in Scotland (where he received an MBA degree and a M.Sc. degree). After working in Canadian industry for far too many years, Malcolm followed his passion for the financial markets by becoming an Investment Advisor/Commodity Trading Advisor with an independent brokerage firm in western Canada. Today, he resides in Saskatchewan, Canada where he trades the financial markets using technical chart analysis, esoteric mathematics and the planetary principles outlined in this book.

Malcolm is the author of several books. His first book, The Bull, the Bear and the Planets, offers the reader an introduction to financial astrology and makes the case that there are esoteric and astrological phenomena that influence the financial markets. His second book, The Lost Science, takes the reader on a deeper journey into planetary events and unique mathematical phenomena that influence financial markets. His third book, De-Mystifying the McWhirter Theory of Stock Market Forecasting seeks to simplify and illustrate the McWhirter methodology. *The Cosmic Clock* follows from the *Lost Science* and helps the reader become better acquainted with planetary events that influence markets. Malcolm has been writing the *Financial Astrology Almanac* each year since 2014.

Malcolm maintains a website (www.investingsuccess.ca) where he provides traders and investors with astrological insights into the financial markets. He also offers the *Astrology Letter* service where subscribers receive twice-monthly previews of pending astrological events that stand to influence markets. He also offers the *Cycle Report* where subscribers are kept apprised on cyclical turning points based on fractals and Hurst cycles.

OTHER BOOKS BY THE AUTHOR

The Bull, The Bear and The Planets

Once maligned by many, the subject of financial astrology is now experiencing a revival as traders and investors seek deeper insight into the forces that move the financial markets.

The markets are a dynamic entity fueled by many factors, some of which we can easily comprehend, some of which are esoteric. *The Bull, The Bear and the Planets* introduces the reader to the notion that astrological phenomena can influence price action on financial markets and create trend changes across both short and longer term time horizons. From an introduction to the historical basics behind astrology through to an examination of lunar astrology and planetary aspects, the numerous illustrated examples in this book will introduce the reader to the power of astrology and its impact on both equity markets and commodity futures markets.

The Lost Science

The financial markets are a reflection of the psychological emotions of traders and investors. These emotions ebb and flow in harmony with the forces of nature.

Scientific techniques and phenomena such as square root mathematics, the Golden Mean, the Golden Sequence, lunar events, planetary transits and planetary aspects have been used by civilizations dating as far back as the ancient Egyptians in order to comprehend the forces of nature.

The emotions of traders and investors can fluctuate in accordance with these forces of nature. Lunar events can be seen to align with trend changes on financial markets. Significant market cycles align with planetary transits and aspects. Price patterns on stocks, commodity futures and market indices can be seen to conform to square root and Golden Mean mathematics.

In the early years of the 20th century the most successful traders on Wall Street, including the venerable W.D. Gann, used these scientific techniques and phenomena to profit from the markets. However, over the ensuing decades as technology has advanced, the science has been lost.

The Lost Science acquaints the reader with an extensive range of astrological and mathematical phenomena. From the Golden Mean and Fibonacci Sequence, to planetary transit lines and square roots through to an examination of lunar and planetary aspects, the numerous illustrated examples in this book show the reader how these unique scientific phenomena impact the financial markets.

Stock Market Forecasting: The McWhirter Method De-Mystified

Stock Market Forecasting - The McWhirter Method De-Mystified

M.G. Bucholtz, B.Sc., MBA

Very little is known about Louise McWhirter, except that in 1937 she wrote the book, *McWhirter Theory of Stock Market Forecasting*.

In my travels to places as far away as the British Library in London, England to research financial Astrology, not once did I come across any other books by her. Not once did I find any other book from her era that even mentioned her name. I find all of this to be deeply mysterious. Whoever she was, she wrote only one book. It is a powerful one that is as accurate today as it was back in 1937. The purpose of writing this book is suggested by the title itself — to de-mystify McWhirter's methodology.

The Cosmic Clock

THE COSMIC CLOCK
TIMING THE FINANCIAL
MARKETS USING
THE PLANETS

M.G. Bucholtz, B.Sc, MBA

Can the movements of the Moon affect the stock market?

Are price swings on Crude Oil, Soybeans, the British pound and other financial instruments a reflection of planetary placements?

The answer to these questions is YES. Changes in price trends on the markets are in fact related to our changing emotions. Our emotions, in turn, are impacted by the changing events in our cosmos.

In the early part of the 20th century, many successful traders on Wall Street, including the venerable W.D. Gann and the mysterious Louise

McWhirter, understood that emotion was linked to the forces of the cosmos. They used astrological events and esoteric mathematics to predict changes in price trend and to profit from the markets.

However, by the latter part of the 20th century, the investment community had become more comfortable in relying on academic financial theory and the opinions of colorful television media personalities, all wrapped up in a buy and hold mentality.

The Cosmic Clock has been written for traders and investors who are seeking to gain an understanding of the cosmic forces that influence emotion and the financial markets.

This book will acquaint you with an extensive range of astrological and mathematical phenomena. From the Golden Mean and Fibonacci Sequence through planetary transit lines, quantum lines, the McWhirter method, planetary conjunctions and market cycles. The numerous illustrated examples show how these unique phenomena can deepen your understanding of the financial markets with the goal of making you a better trader and investor.

Made in the USA
Las Vegas, NV
21 August 2023